CONSUMER BEHAVIOR AND ENERGY CONSERVATION

Consumer Behavior and Energy Conservation

A Policy-Oriented Experimental Field Study on the Effectiveness of Behavioral Interventions Promoting Residential Energy Conservation

by

Peter Ester

1985 **MARTINUS NIJHOFF PUBLISHERS**
a member of the KLUWER ACADEMIC PUBLISHERS GROUP
DORDRECHT / BOSTON / LANCASTER

Distributors

for the United States and Canada: Kluwer Academic Publishers, 190 Old Derby Street, Hingham, MA 02043, USA
for the UK and Ireland: Kluwer Academic Publishers, MTP Press Limited, Falcon House, Queen Square, Lancaster LA1 1RN, UK
for all other countries: Kluwer Academic Publishers Group, Distribution Center, P.O. Box 322, 3300 AH Dordrecht, The Netherlands

Library of Congress Cataloging in Publication Data

ISBN 90-247-3134-8

PRINTED IN THE NETHERLANDS

CONTENTS

PART III RESULTS

Acknowledgements

This study was supported by a grant from the Dutch Ministry of Economic Affairs for which I would like to express my sincere gratitude. A scientific advisory committee actively assisted in designing this study. I greatly acknowledge the participation and expertise of the following members: Anne W. van den Ban (Agricultural University, Wageningen), Cees J.H. Midden (Energy Study Center/ Netherlands Energy Research Foundation), W. Fred van Raaij (Erasmus University, Rotterdam), Gerard M. Rosinga (Ministry of Economic Affairs), and Wouter J. Veening (Ministry of Public Housing, Physical Planning and Environmental Protection). In addition, I would like to thank my promotors Anne W. van den Ban and W. Fred van Raaij for carefully reviewing earlier drafts of this book. I learned much from their expertise in the area of, respectively, communications theory and economic psychology.

My former colleague Joop de Boer (Free University, Amsterdam) actively assisted in almost all phases of this study, including data collection and data analysis, for which I am - more than words can express - most grateful. Also, I would like to thank Lou J.M. Schreurs who was involved at the start of this project and Cees Vermeulen for his participation as student-assistent. Next, I greatly acknowledge the help of Rob Quint (Dutch Foundation for Energy Conservation Information) in compiling the conservation information leaflet which was tested in this study. Veldkamp Marketing Research, Inc. (Amsterdam) carried out the actual interviewing and coding of the data. Their work was well beyond the requirements of a formal agreement.

I am especially grateful to the hundreds of households which participated in this study, whose indispensable co-operation is usually played down in behavioral research by little respectfully calling them "subjects".

This study greatly benefited from the meter reading assistance of the utility companies in the five research locations: PEN, Gemeenschappelijk Gasbedrijf Kop Noord-Holland, GEB/GZW-Zaandam, GEB-Amsterdam, and GEB-Hilversum.

Next, I would like to thank the Department of Psychology, Claremont Graduate School, California and the Department of Psychology, Virginia Polytechnic Institute and State University, Blacksburg, Virginia, for offering me the opportunity to work on this book for a number of months. This visit was made possible by a Fulbright grant for which I am most grateful. I greatly appreciated the stimulating discussions with John G. Cope, E. Scott Geller, Mark W. Lipsey,

Keith A. Neuman, Stuart Oskamp, Daniel Stokols, Alan W. Wicker, and Richard A. Winett. I also would like to thank the Institute for Environmental Studies (Free University of Amsterdam) for permitting this sabbathical leave.
A number of "foreign" colleagues took the trouble of commenting upon an earlier draft of this manuscript. In this respect I would like to thank George Gaskell (Department of Social Psychology, London School of Economics and Political Science), Paul C. Stern (National Academy of Sciences), and Richard A. Winett (Department of Psychology, Virginia Polytechnic Institute and State University). With all this help, I am nevertheless responsible for the design, analysis, and conclusions of this study.
I am indebted to Carol Mindell for correcting and improving the English text, and to Karin George for the excellent and dedicated way in which she took care of typing several versions of this manuscript.
Last but not least I want to thank my wife Carrie for many things, but most of all for sharing my life.

Zeist/The Netherlands, November 1984
Peter Ester

LIST OF TABLES

LIST OF FIGURES

ORGANIZATION OF THIS BOOK

The primary goal of this study is to investigate the effectiveness of behavioral interventions aimed at promoting energy conservation by consumers. Part I of the book will present some theoretical reflections on this subject. Chapter 1 contains a general introduction to the main theme of this study. It is argued that energy conservation has many advantages over other energy policy options, especially with respect to sociopolitical, economic, technical, safety, and environmental considerations. Particular emphasis is placed on behavioral aspects of consumer energy conservation, which illustrate the importance of a social scientific approach to energy conservation. Next, the research questions underlying this study are described. In Chapter 2 some theories are discussed which deal with micromotives behind individual energy consumption and their - often unintended - macroconsequences. Also, some personal correlates of consumer energy conserving behavior will be examined, including general energy attitudes, energy knowledge, and consumers' lifestyles and sociodemographic characteristics. Chapter 3 analyzes some antecedent and consequence behavioral interventions (information/education, prompting, modeling, feedback, self-monitoring, reinforcement/punishment) for promoting energy conservation by consumers and presents an extensive review of behavioral experiments conducted in this area. Major shortcomings of these experiments are revealed, both with respect to methodological criteria and policy relevance. Chapter 4 shows how an attempt is made in this field experimental study to overcome these shortcomings. The following interventions are selected to be tested in this study: energy conservation information, biweekly and monthly energy consumption feedback, and self-monitoring of energy consumption. Next, the research hypotheses are formulated which guided this study.

Part II contains a description of the experimental design of the study. Chapter 5 discusses how this field experiment, which was conducted in five cities in the Netherlands with approximately 400 subjects, has been implemented. Information will be provided with respect to research locations, target group, sampling procedures, recruitment of subjects, experimental conditions, group assignment, baseline energy consumption calculations, correction factors, pre- and post-experimental interviews with subjects, nonresponse, and post-experimental meter readings. Finally, a graphic representation of the main research variables will be included.

Part III of this book turns to the empirical results of the field experiment. In Chapter 6 some hypotheses will be tested regarding belief structures with respect to energy conservation. An attempt will be made to explain consumers' intentions to conserve energy in the household by beliefs about consequences of energy conservation, evaluations of these consequences, normative beliefs about energy conservation, and compliance with these normative beliefs. Also, the role of consumers' knowledge of residential energy matters will be examined. Chapter 7 provides data about the absolute and relative effectiveness of energy conservation information, biweekly and monthly feedback, and self-monitoring in encouraging consumers to conserve energy. Effectiveness of these behavioral interventions will be determined by examining actual energy consumption measures. In Chapter 8 some hypotheses will be tested about cognitive aspects related to responsiveness to behavioral interventions and attitude change toward energy conservation.

Part IV summarizes the main findings and draws some conclusions. Chapter 9, presents a number of conclusions from this study and formulates some energy policy recommendations as well as directions for future research. Chapter 10, finally, contains a summary of the main outcomes of this study.

PART I THEORETICAL CONSIDERATIONS

1. THE NEED FOR ENERGY CONSERVATION: INTRODUCTION TO THEME AND BACKGROUND OF THIS STUDY

In the past decade it has become alarmingly apparent that the world's oil and natural gas resources are not only finite but also diminishing at a significant pace. Although there is no wide consensus on the exact reserves of these nonrenewable resources (Odell & Rosing, 1980), there is little disagreement over the fact that the rate of depletion will profoundly affect the lifestyle of future generations (National Academy of Sciences, 1979, 1980; Stern & Gardner, 1980; Stobaugh & Yergin, 1979).

Whether this will consequently imply lowering the _quality_ of life of large segments of the population is still debated and depends above all on policy choices that have to be made between different energy options and strategies, as well as on available energy technologies.

For a comprehensive understanding and analysis of the global energy situation one has to add at once that depletion of nonrenewable fossils is not the only factor which causes observers to speak of an energy question or even of an energy "crisis". Among the factors mentioned are: the unequal geographical distribution of energy resources, the disproportional energy consumption by and between developed and developing countries, social and economic impacts of rising energy prices, the use of oil supply limitations by oil producing countries to intensify international political pressure, temporary acute energy shortages, the immense waste of energy in industrialized countries, economic, technical, environmental and social constraints with regards to the development and use of energy resources other than oil and natural gas (e.g. nuclear energy, coal, solar energy), and institutional and structural barriers with respect to the adoption and diffusion of innovative energy technologies and policies.

The complexity of the energy question is clearly reflected by the diversity and heterogeneity of the above-mentioned factors.

1.1 Energy transition

In most energy scenarios the year 2000 plays a crucial role. By that year it is believed the necessary transition should be completed from a predominantly oil-dependent society to a society which utilizes energy sources that are either renewable or available on a scale large enough for centuries. If not, serious energy gaps are expected at the end of this century as oil supply will probably no longer meet its demand (cf. CONAES, 1979; Institute for Energy Analysis, 1976; International Institute for Applied Systems Analysis, 1981; MOPPS, 1977;

Resources for the Future, 1979; WAES, 1977).
Though already dated - which seems to be the fate of many energy scenarios - the WAES scenario sketches the energy situation for the remaining part of this century as follows: "Energy supply and demand, when taken together over the period 1971-2000, paint a disconcerting picture: growing shortages of oil. (...) Resource and production limitations begin to restrict oil supply in the period 1985-2000. Further increases in oil demand beyond 1990 must be satisfied from other fuels. The prospective oil shortfall must be filled or eliminated. Oil demand and supply must always balance" (WAES, 1977, p. 235).
Thus, transition from nonrenewable to renewable energy resources seems to be a fundamental task for any intelligent energy policy. However, "the problem is in effecting a socially acceptable and smooth transition from gradually depleting resources of oil and natural gas to new technologies whose potentials are not now fully developed or assessed and whose costs are generally unpredictable. This transition involves time for planning and development on the scale of half a century. The question is whether we are diligent, clever, and lucky enough to make this inevitable transistion an orderly and smooth one" (CONAES, 1979, p. 72).
The crucial problem is, of course, that almost all existing or proposed technologies and energy options which in principle come into consideration for this necessary transition of energy resources are characterized by some sort of negative economic, social, environmental, technical and/or political side effects. A short look at some energy options may illustrate this problem.

1.1.1 Coal

Coal, which as such is abundantly available (1), used as a substitute for oil and natural gas involves tremendous financial investments, requires a specific and presently often lacking infrastructure, and causes significant transport and storage problems. But above all, coal produces unprecedented amounts of air pollution (sulfur dioxide, nitrogen dioxide, carbon oxides, particulate matter, heavy metals) and thus unforeseen health effects. In its study on global energy supply options for the period 1980-2030 the International Institute for Applied Systems Analysis concludes that: "Coal should not be used as the major source of primary energy for meeting world demand for large amounts of energy over the next fifty years, since this would not only deplete its resource base within one hundred years but also create severe environmental and human health hazards" (International Institute for Applied Systems Analysis, 1981, p. 28).

1.1.2 Nuclear energy

Nuclear energy poses a different set of problems centered around five issues in particular: the safety of routine operation of the nuclear fuel cycle, the possibility of severe nuclear accidents, the unsolved question of storing nuclear waste, possible production of nuclear bombs by countries that did not sign the Non-Proliferation Treaty or by terrorist groups, and the almost inevitable necessity of having a centralized-hierarchical power structure to control the nuclear energy system. This enumeration indicates that the nuclear debate is not just a controversy over strictly technical problems and solutions, but more so over major social-political dimensions and impacts (Mazur, 1975; Kasperson, Berk, Pijawka, Sharaf & Wood, 1980).
In correspondence with this twofold controversy a number of social science studies have indicated that proponents and opponents of nuclear energy differ not only in their subjective estimates of the likelihood of nuclear accidents, but also in their evaluation of social, economic and environmental consequences of nuclear energy (see e.g. Eiser & van der Pligt, 1980, Ester, van der Linden & van der Pligt, 1982; van der Linden, Ester & van der Pligt, 1982; Otway & Fishbein, 1976, 1977; Otway, Maurer & Thomas, 1978; Slovic, Fischhoff & Lichtenstein, 1981; Stallen & Meertens, 1979, 1981; Thomas, Maurer, Fishbein, Otway, Hinkle & Simpson, 1978; Woo & Castore, 1980).
Public acceptance thus seems to be essential for nuclear energy to become a viable option for energy policy. There is, however, strong empirical evidence that the opposition to nuclear energy within Western industrialized countries is growing, not only in number but also in magnitude (cf. Abma, Jägers & van Kempen, 1981; de Boer, 1977, Ester, de Rooij & Schreurs, 1979; Melber, Nealy, Hammersla & Rankin, 1977). This growing opposition - embodied by the environmental movements in these countries - has urged some governments to delay the development of nuclear energy programs.
In the Netherlands, for instance, the government decided to organize and finance a public inquiry on energy policy options and choices to enable the Dutch people to participate in a large-scale public debate on energy policy in general and on nuclear energy in particular. (2) The debate was structured through public hearings, discussion meetings, information exchange by proponents and opponents of nuclear energy, and lasted for two years. Total costs of this public inquiry has been estimated at twenty-eight million guilders.
It had been agreed that no formal decision on the intended expansion of the nuclear energy program will be taken before the outcomes of this unique social policy experiment became known.
Thus, apart from technical and economic considerations, there are strong social

constraints with regards to increasing the contribution of nuclear power for future energy needs. Also, there is considerable doubt - contrary to conventional wisdom - on the extent to which nuclear energy will be able to contribute significantly to energy supply in the short and medium-long run. Bupp (1979) concludes from his analysis of the conceivable role of nuclear power in the U.S. energy future: "In the United States there is simply no reasonable possibility for 'massive contributions' from nuclear power for at least the rest of the twentieth century (...) nuclear power offers no solution to the problems of America's growing dependence on imported oil for the rest of this century" (Bupp, 1979, p. 135). This picture changes, of course, if one lookes at the contributions from nuclear power in the long run (e.g. breeder reactors).

1.1.3 Renewable sources

Renewable sources constitute a third set of possible energy policy options. The main advantage of renewables (e.g. hard and soft solar technology, biomass, hydroelectricity, wind energy, geothermal energy, ocean thermal energy conversion, tidal power) obviously is the fact that they are available on an inexhaustible basis.

In addition to this, one has to mention the overwhelming fact that they are clean (though not necessarily in the production phase) and sustainable. Nevertheless, most observers do not expect the contribution of renewable sources to be very significant under present conditions before the year 2000. On the other hand, however, they agree that the possibility of major breakthroughs of either a technological or cost nature could drastically alter this prospect.

Apart from the technical obstacles (e.g. solar energy storage) the weak competitive position of most renewables compared with other available technologies is also due to socioeconomic and institutional barriers (cf. Maidique, 1979). First of all, capital costs of developing renewable sources are very high which, under present market conditions and price comparisons, limit their commercial use, at least on a large-scale basis. This will of course change if costs of traditional energy supplies rise.

Also, particularly in a situation of economic uncertainty fast paybacks are expected from high capital investments. Social science studies on the adoption of solar energy equipment by individual consumers, for instance, indicate that the expected payback period is an important determinant of adopting solar energy (Berkowitz & Haines, Jr., 1980; Gerlach, Renz & Brown, 1979; Klein, 1979; Krusche, 1979; Labay & Kinnear, 1980; Leonard-Barton, 1980; Schleyer & Young, 1977, Sparrow, Warkov & Kass, 1980; van Raaij, 1981; Warkov, 1980; Weiss,

1979). (3)
Ultimately, a sensible macro-evaluation of energy options should compare capital costs of developing renewable sources with the capital costs of transition from a basically oil-dependent energy system to a coal- and/or nuclear energy system, including externality costs (pollution, risk management). In this respect one has to realize that a notable increase in the use of renewables results in a gradual decrease of its capital costs. Also, one has to take into consideration that due to an imbalance - or at least divergent priorities - in research and development funding, nuclear and coal technology is more advanced than technologies regarding renewable sources.
This last factor indicates that institutional barriers are important for understanding the present state-of-affairs of renewables technology.
According to Maidique (1979) these barriers must be overcome if renewable sources are to have a fair chance in the marketplace against conventional sources. As an example he mentions solar heating which is far less diffused than possible in the United States as a consequence of institutional barriers such as inaccurate building codes, lack of skills in installing and maintaining solar systems, passive attitude and role of utilities, and lack of economic incentives for adopting solar heating. Maidique projects that by overcoming these institutional barriers and given reasonable incentives, solar energy could provide between a fifth and a quarter of the U.S. energy needs by the turn of this century (Maidique, 1979, p. 183).
Geographical and climatological conditions are evidently of vital importance for predicting the contribution of solar energy and other renewable sources in other countries. In the Netherlands, for instance, governmental reports estimate this contribution between 1.5% and 1.9% by the year 2000 (Nota Energiebeleid, deel I, 1979).

1.1.4 Conservation

Though this overview of various constraints of a number of relevant energy options is by no means meant to be exhaustive, it nevertheless clarifies and illustrates the former statement that each energy option is confronted with some sort of sociopolitical, economic, technical, safety or environmental difficulties.
However, thus far we have not spoken about an energy source which as such is quite promising and hardly controversial and which has been labeled by some observers as "the key energy source" (Yergin, 1979) - namely, conservation. It is almost a cliché by now to say that a huge amount of energy is being wasted

in modern industrialized countries. Still, the proportions of energy waste justify continuous attention to this problem as well as to solutions to it (cf. Darmstadter, Dunkerley & Alterman, 1977; Foley, 1976; Griffin, 1979; Schipper & Lichtenberg, 1977; Yergin, 1979).

This is especially true for the United States and for Canada, since their energy consumption per dollar of gross domestic product is much higher than any other industrial nation. West Germany, for instance consumed in 1976 less than three quarters as much energy as the United States for each dollar of gross domestic product, and France only half (Darmstadter, Dunkerley & Alterman, 1977). Although Sweden and Canada have approximately the same gross domestic product per capita, Canada consumes on average twice as much energy (Foley, 1976). Also, Sweden uses roughly 60 percent less energy as the United States to generate each dollar of gross national product (Schipper & Lichtenberg, 1976). Although caution is needed in interpreting such intercountry energy consumption comparisons, they indicate that substantial energy savings can be reached in the United States and Canada. It is not surprising therefore that, given the outcomes of these comparisons, several observers have concluded that the U.S. energy consumption can be reduced considerable <u>without</u> major lifestyle impacts. "If the United States were to make a serious commitment to conservation, it might well consume 30 to 40 percent less energy than it does now, and still enjoy the same or an even higher standard of living" (Yergin, 1979, p. 136).

Other energy analysts have calculated similar conservation percentages (cf. Joerges & Olsen, 1979; National Academy of Sciences, 1980; Olsen & Joerges, 1981; Ross & Williams, 1978).

Though it is a well-established fact that European countries have notably lower levels of energy consumption than the United States and Canada, recent forecasts have nevertheless shown that stringent conservation measures still can reduce energy consumption considerably in those countries (Barth, 1981; Leach, 1979; Meyer-Abich, 1979; Potma, 1979).

It is therefore not surprising that conservation is a major cornerstone of energy policies of many industrial societies. The advantages of energy conservation over other energy options are concisely summarized in the words of the following author: "conservation may well be the cheapest, safest, most productive energy alternative readily available in large amounts. By comparison, conservation is a quality energy source. It does not threaten to undermine the international monetary system, nor does it emit carbon dioxide into the atmosphere, nor does it generate problems comparable to nuclear waste. And contrary to the conventional wisdom, conservation can stimulate innovation, employment, and economic growth" (Yergin, 1979, p. 137).

Taking the Dutch energy policy as an example (see de Boer, Ester, Mindell &

Schopman, 1982, a + b), the following conservation target figures per energy-using sector have been fixed by the government for the 1977-2000 period.

Table 1.1: Conservation Target Figures per Energy-Using Sector in the Netherlands, 1977-2000

Sector	Percentage of Total Domestic Consumption	Efficiency Improvement 1977-2000
1. Residential space heating	15%	45%
2. Other residential energy consumption	7%	20%
3. Utility- and service buildings	18%	35%
4. Fuel consumption in transportation sector	13%	25%
5. Industrial energy consumption	47%	30%
Total domestic energy consumption	100%	30%

Note. From Nota Energiebeleid, deel I. 's-Gravenhage: Staatsuitgeverij, 1979, p. 72.

As Table 1.1 indicates the intended overall energy efficiency improvement figure for the 1977-2000 period is 30 percent: 45 percent efficiency improvement has to be realized in the residential space heating sector, 20 percent in other residential energy consumption (e.g. electricity), 35 percent in utility- and service buildings (service sector, agriculture, government), 25 percent in the transportation sector, and finally, a 30 percent energy efficiency improvement will have to be achieved in industry.

In order to reach those target figures conservation programs have been set up for each of these energy sectors. Total costs of these conservation programs are estimated at 63 billion Dutch guilders: 5 billion will be invested in energy-efficient durable consumer goods, 19 billion in the housing sector and 39 billion guilders in industry.

Proposed conservation policy measures include among others: a national insulation program which aims through subsidies and other facilities at insulating 200,000 houses each year, a special conservation program for the government in its role as an energy consumer as well as for the non-profit sector, implementation of new regulations (e.g. standard insulation norms for new buildings and houses, heating efficiency norms), measures encouraging selective car use, energy labels on household appliances, energy audits, energy taxes, changes in energy rate structures, energy efficiency demonstration projects, national research programs on energy conservation.

As mentioned before, one of the most important advantages of energy conservation obviously is that conservation is far less controversial as compared with energy options like, for instance, coal and nuclear power. This is mainly due to the fact that there are no major environmental, safety or economic constraints.

Although this is not a study on environmental and economic impacts of energy conservation the following two examples of environmental and macro-economic consequences of conservation might nevertheless be illustrative.

In Table 1.1 conservation target figures were summarized for each energy-using sector in the Netherlands as formulated by the Dutch energy policy. Proposed energy conservation measures to reach these target figures were mentioned. Table 1.2 gives a rough estimate of avoided air pollution in the Netherlands in the 1985-2000 period as a consequence of these policy measures as far as conservation of oil products in concerned. Estimates are based on the present state of environmental technology.

Table 1.2: Broad Estimate of Avoided Air Pollution in the Netherlands Resulting From Conservation of Oil Products*

	1985	1990	2000
Carbon monoxide	7%	10%	13%
Nitrogen oxides	9%	16%	28%
Sulfur dioxide	22%	36%	56%
Aldehydes	26%	40%	58%
Hydrocarbons	7%	11%	15%
Particulate matter	18%	30%	48%

* in percentage of expected pollution from oil consumption without conservation

Note. From Nota Energiebeleid, deel I. 's-Gravenhage: Staatsuitgeverij, 1979, p. 102

Table 1.2 cleary shows that air pollution can be reduced - or better: avoided - considerably through conservation of energy. This is particularly true for aldehydes, sulfur dioxide, and particular matter and to a lesser degree for pollutants like nitrogen oxides, hydrocarbons, and carbon monoxide. Thus, a conservation-oriented energy policy can significantly contribute to a preventive environmental policy.
Like the development of any major energy resource, a stringent conservation policy requires enormous investments. But conservation also generates economic benefits, not only in the micro-economic sense of a lower energy bill but also in the macro-economic connotation of positive employment effects.
If we take the Dutch National Insulation Program as an example, which aims through financial incentives to insulate 2.5 million houses between 1978 and 1990, a total government investment of 1.3 billion Dutch guilders is needed. A recent study by Bruggink (1981) has shown that the employment benefits of the program amount to 51,000 personyears. (4)

1.2 The importance of energy behavior in understanding energy consumption

As indicated above, there are obviously very strong arguments in favor of energy conservation. Traditionally, governmental conservation policy is primarily directed at promoting technological energy efficiency improvements (e.g. insulation, more fuel efficient automobiles). Conservation seems to be conceptualized in the first place as a technological problem which consequently requires a technological solution (Darley, 1978).
This conceptualization is referred to as the so-called "technological fix" (Heberlein, 1974). One of the main starting points of this study, however, is that energy conservation should not just be defined as a predominantly technological problem, but also needs to be recognized as fundamentally a behavioral problem (Baum & Singer, 1981; Cone & Hayes, 1980; Geller, Winett & Everett, 1982; Lipsey, 1977; Olsen & Joerges, 1981; Seligman & Becker, 1981).
This is true because energy consumption in general and energy conservation in particular are related to human behavior, lifestyles, values and attitudes (Oskamp, 1981). Therefore, it is somewhat naive to think that the sole availability of energy efficient technological innovations as such is a sufficient condition for energy conservation, for innovations have to be adopted (Darley, 1978; Darley & Beniger, 1981; van Raaij, 1981). Diffusion of innovation theory suggests that besides factual availability, adoption of an innovation is a function of a number of adopter- and innovation-related characteristics (Robertson, 1971; Rogers & Shoemaker, 1971). (5) In short: "While technological

innovations (...) are an effective means of significantly reducing consumption, by ignoring social systems and making simplistic assumptions about human nature, sole reliance on this approach represents at best a partial solution" (Neuman, 1980, p. 2).

Several studies have provided empirical evidence for the notion that human behavior is an important factor in explaining and understanding energy consumption levels. Wotaki (1977) observed in a study in New Jersey that even after houses had been insulated the variance in energy consumption remained almost the same compared to the situation prior to the insulation took place (cf. Socolow, 1978).

In his study on the variability of energy consumption in nominally similar houses, Sonderegger (1978) found that 54% of the observed variance is explained by conventional physical housing characteristics (e.g. number of bedrooms). It appeared that 71% of the remaining 46% is explained by resident-dependent factors. (6)

Verhallen and van Raaij (1979, 1981) concluded from their study of determinants of energy consumption (heating) in a sample of similar houses in a town (Vlaardingen) in the Netherlands that 26% of the variance in consumption is explained by differences in behavior of residents. More specifically they found six factors in energy-related household behavior, plus two additional factors: bedroom temperature while sleeping, home temperature during absence, use of window curtains, airing of rooms, use of bedrooms for studying/playing, use of halldoor and use of pilot flame. (7) These authors make a distiction between purchase-related and usage-related energy behavior. Purchase-related behavior refers to the purchase of appliances and home improvements (e.g. insulation, retrofitting, double glazing), whereas usage-related behavior refers to changes in daily behavioral routines (e.g. lowering the thermostat at night, using fewer rooms, closing curtains). (8)

Another, quite different, facet which underlines the importance of a behavioral approach to energy conservation is the fact that for an energy conservation policy to be effective it has to be accepted by the target group. Thus, a 55 mph speed limit is not very effective if motorists do not accept this limit. In other words, conservation policy acceptance is a necessary condition for conservation policy effectiveness (Joerges & Olsen, 1979; Leeuw & Ester, 1981; Morrison, 1978; Olsen, 1981; Olsen & Goodnight, 1977).

Theoretical and empirical evidence suggests that (energy) policy acceptance is a function of at least three factors: the degree in which individual freedoms are affected, the degree of social control, as well as possibilities to elude social control (Ester, 1979a).

1.3 Energy conservation and the behavioral sciences

The fact that energy conservation has a crucial behavioral dimension besides a technological component implies an important role for the behavioral sciences (Wilbanks, 1981). Because if it is true that human behavior is an essential factor in explaining energy consumption, then changing human behavior will be a significant contribution to energy conservation (Cone & Hayes, 1980; Geller, Winett & Everett, 1982). According to Maloney and Ward (1973) the ecological crisis is first of all a crisis of maladaptive behavior and therefore "... the most feasible solution lies in the immediate changing of critical behaviors on a population-wide basis" (Maloney & Ward, 1973, p. 583-584).

This last statement indicates that also from a policy point of view a behavioral science perspective on energy conservation, being one solution to the ecological crisis referred to by Maloney and Ward, is simply indispensable. For not only can this perspective offer a valuable contribution to our understanding of energy matters in general and energy conservation in particular, but also theoretical and practical principles and guidelines can be derived as to how these "critical behaviors" can be changed in view of the urgent need for energy conservation.

If one looks at energy conservation research as a whole one is forced to conclude, however, that the dominant research focus is on technological innovations, whereas there are only relatively few studies on behavioral innovations (Wilbanks, 1981). This is not just true for energy research but also for environmental research in general (Ester & Leroy, 1984; Hoogerwerf, 1980).

It has been estimated from an ongoing registration of research projects in the Netherlands on environmental issues that less than 5 percent of these projects is on social aspects of these issues, and less than half a percent on societal causes of environmental problems (Nelissen, 1977). Recent developments, fortunately, show considerable progress (LaSOM, 1979, 1981). These figures do not, of course, only reflect policy makers' (un)willingness to subsidize behavioral environmental and energy research but in a way also indicate the existing priorities on the research agendas of social scientists. According to some observers this last aspect is especially the case with sociologists: "a scientist investigating relationships between energy and society can expect little help from the literature of sociology published in the last 20 years" (Duncan, 1978, p. 1). The pioneering work of Cottrell (1955) can be characterized as a notable exception.

Several reasons have recently been suggested for the puzzling fact that sociology has largely ignored the relationship between human societies and their biophysical environments (Catton, Jr. & Dunlap, 1978, 1980; Dunlap, 1980;

Dunlap & Catton, Jr., 1979a, 1979b).
One of the reasons put forward is that "the discipline of sociology is premised on a set of background assumptions or a paradigm that has led sociologists - regardless of their particular theoretical persuasion - to treat human societies as _if_ they were exempt from ecological constraints. As part of their emphasis on the exceptional characteristics of humans, most sociologists have totally ignored the biophysical environment, as if human societies somehow no longer depend on it for their physical excistence and for the means of pursuing the goals they value" (Catton, Jr. & Dunlap, 1980, p. 25). (9)
One of these background assumptions is the well-known and influential but also almost "anthropocentric" Durkheimian methodological rule that in sociology social facts are only to be explained by other social facts and not by non-social modalities (Durkheim, 1895). This might have led sociologists to have little interest in the biophysical environment as such and in ecological constraints to human societies. As a consequence "environmental sociology" is a very poorly developed discipline.
It is interesting to see that the sister discipline of environmental sociology, namely _environmental psychology_, is a relatively advanced area of scientific knowledge (Stokols, 1978). Though many definitions of this discipline circulate one might define environmental psychology as "the study of the interrelationship between behavior and the built and natural environment" (Bell, Fisher & Loomis, 1978, p. 6). Topical areas within environmental psychology include, for example, human spatial behavior, subjective responses to environmental stressors (pollution, noise, crowding), environmental cognition, cognitive mapping, environmental attitudes and environmental behavior, behavioral interventions promoting ecologically conscious behavior (cf. Altman, 1975; Altman & Wohlwill, 1976, 1977; Craik & Zube, 1976; Heimstra & McFarling, 1974; Moos & Insel, 1974; Proshansky, Ittelson & Rivlin, 1976).
Given the relevance of especially the last two topics for the subject of this study, later chapters will extensively lean on major theoretical and empirical findings from environmental psychology.

Summarizing, thus far it has been argued that energy conservation is an indispensable and, in theory, quite feasible energy option. Because of the fact that energy consumption directly or indirectly is related to people's attitudes, values, norms, lifestyles, and personal contexts and circumstances, it has been argued that major characteristics of the energy question are not just of a technological but also of a _social_ nature. Therefore, besides a technological approach, a social science reflection on both causes of and solutions to the energy question is needed. For energy conservation - being one of these "solutions" - this implies that policy makers cannot solely rely on technological innovations but also need to promote behavioral innovations. However, the question arises: but how? This question is especially relevant given the fact that social scientists, with the exception of environmental psychologists, have been relatively passive with answering this urgent and highly policy relevant question.

To be more specific: which policy instruments - or in more formal language: which behavioral interventions - could policy makers implement to promote pro-energy conservation attitudes and lifestyles? What is the scientific status of theories about attitude and behavior change underlying these interventions? What is the absolute and relative effectiveness of these interventions? What are their short-term and long-term social impacts? Are there any positive or negative unintended side-effects? How will the policy target group respond to these interventions? Are there differences to be expected in responsiveness between different segments within the target group? How cost-effective are various behavioral interventions?

It might be clarifying to illustrate the relevance of these questions by applying them to the subject of this study: the theoretical and empirical effectiveness of behavioral interventions in modifying consumers' attitudes and behaviors regarding energy conservation.

As will be shown in Chapter 3, a range of behavioral interventions can be implemented to promote energy conservation by consumers (e.g. information, education, prompts, modeling, feedback, self-monitoring, rewards, punishments). The first question a policy maker is faced with is: what is the relative effectiveness of these interventions (e.g. is rewarding consumers for reduced energy consumption more or less effective than punishing them financially for excessive consumption?)? How persistent will this attitude and behavior change eventually be (short-term or long-term?)? Is it to be expected that any of these interventions used as energy conservation policy instruments will yield unintended side-effects (e.g. will different income groups respond differently to financial stimuli for energy conservation?)? And last but not least: how cost-effective are these interventions (e.g. what is the relationship between costs of these interventions and obtained benefits in terms of the amount of energy saved?)?

These and similar questions have to be answered before implementing these interventions on a large scale basis (Cone & Hayes, 1980; Ester, 1979a; Geller, Winett & Everett, 1982; Olsen & Joerges, 1981; Seligman & Hutton, 1981).

1.4 Specification of research problems

By now, it is possible to define the main research problems of this study more precisely. Subsequent chapters will offer further specification.

(1) which factors regulate the (un)willingness of consumers to conserve energy in their households?

(2) which behavioral interventions can be used to promote energy conservation by consumers and what is their estimated effectiveness (both in terms of behavior change and energy savings) from a theoretical point of view?

(3) what is the actual or empirical effectiveness of a number of these interventions?

ad 1. In order to answer this research problem a number of theoretical explanations will be outlined of the micro-motives behind the willingness or unwillingness of consumers to reduce their energy consumption. In other words, an attempt will be made to specify some determinants of a consumer's intention or decision to conserve energy.

ad 2. A number of behavioral interventions will be described which either are presently used by policy makers to encourage energy conservation in the consumer sector or which could be implemented in the future. Based on existing social science know-how the theoretical validity of underlying assumptions about attitude and behavior change will be determined.

ad 3. To answer this research question some of these behavioral interventions will be selected to be tested empirically in a field experiment. Subjects are about 400 consumers from five towns in the Netherlands. The experiment lasted for one heating season (1980-1981). The experimental design permits an answer to the question about both the absolute and relative empirical effectiveness of behavioral interventions promoting energy conservation by consumers.
It is obvious that these three research problems are closely related. In fact, they are characterized by a certain "logic of sequence". This means that answering a particular research problem presupposes that adequate solutions to - from a theoretical point of view - logically preceding research problems have been found. (11)
To be more specific, having an adequate answer to the question under which con-

ditions consumers are (un)willing to conserve energy (research problem _one_) is indispensable for establishing both the theoretical and empirical power of behavioral interventions directed at encouraging consumers to conserve energy (research problems _two_ and _three_).

For a correct understanding of the scope of this study it is, finally, important to underline that its main emphasis is primarily directed at promoting short-term energy behavior change and not at stimulating the adoption of and investments in technical conservation innovations. Thus, this study looks at only one class of residential conservation actions.

Notes

1. Coal reserves are estimated to be about 600 billion tons of coal equivalent (tce) and coal resources at some 10,000 billion tce (See World Energy Conference, 1978).
2. This public inquiry, which started in September 1981, had two phases. In phase I an independent steering committee appointed by the government made an inventory of present points of view and beliefs about nuclear energy. After verifying those views and beliefs by defining their plausibility and reliability, the committee summarized its findings in an interim report. One of the functions of this report is to inform the general public in an accessible way and on a large scale about all relevant aspects of nuclear energy.

   In phase II the steering committee activated a public discussion on nuclear energy through hearings and special meetings. After this phase, the committee presented the outcomes in a final report to the Dutch parliament and government who then are supposed to define a more definite policy with regards to the question whether or not to increase the number of nuclear power plants in the Netherlands (see Opzet Nota, 1979).
3. See also Cunningham and Joseph (1978), and Hanna (1978).
4. Recently Potma (1979) developed a low energy scenario for the Netherlands which strongy emphasizes conservation. He estimates the positive employment effects of this scenario at 193,000 personyears by the year 2000.
5. See Chapter 2 for a more detailed discussion.
6. Both the study of Wotaki and the research reported by Sonderegger are part of the Twin Rivers project (New Jersey) of a multidisciplinary research group from Princeton University. See Energy and Buildings, 1978, 1 for an overview of results.
7. In a recent study in Zoetermeer (the Netherlands) among 138 residents of identical houses a research team from Delft University found that 35% of the variance in gas consumption was explained by differences in residents' energy-related behaviors (Backer, Bruchem, Hamer, Mellink, Meijer, Moezel, Plugge & Westra, 1980). This outcome may serve as another illustration of the importance of consumer behavior in understanding consumer energy consumption.
8. See also van Raaij (1981)
9. According to Catton, Jr. and Dunlap (1980) the ecological question calls for a fundamental shift in existing sociological paradigms. "The changes ecological conditions confronting human societies seriously challenge sociology,

for the discipline developed in an era when humans seemed exempt from ecological constraints. Disciplinary traditions and assumptions that evolved during the age of exuberant growth imbued sociology with a world view or paradigm which impedes recognition of the societal significance of current ecological realities. Thus, sociology stands in need of a fundamental alteration in its disciplinary paradigm" (Catton, Jr. & Dunlap, 1980, p. 15).

10. Stokols (1978) considers human-environment transaction processes to be the central object of environmental psychology. In his view these processes can be characterized in terms of two basic dimensions: cognitive vs behavioral forms of transaction, and active vs reactive phases of transaction.
    In turn, these two dimensions yield four modes of human-environment transactions:
    a. interpretative mode (active/cognitive), e.g. cognitive mapping.
    b. evaluative mode (reactive/cognitive), e.g. attitudes toward environmental pollution or toward energy conservation.
    c. operative mode (active/behavioral), e.g. experimental analysis of ecologically relevant behavior.
    d. responsive mode (reactive/behavioral), e.g. impact of environmental stressors on subjective well-being.

    This study, which is directed at how consumers perceive energy conservation and how they respond to a number of behavioral interventions aimed at promoting conservation, represents both the evaluative and the operative mode of human-environment transaction.
11. For an interesting application of this "logic of sequence" in social policy research see Fairweather and Tornatzky (1977).

2. CONSUMER ATTITUDES, CONSUMER BEHAVIOR AND ENERGY CONSERVATION: A BEHAVIORAL SCIENCE PERSPECTIVE

2.1 Introduction

Household energy consumption is a function of many and quite different factors. Hypothetically one would expect energy consumption to be related to such divergent variables as comfort preferences, thermostat setting, socioeconomic status, family cycle, energy prices, perceived costs and benefits of energy conservation, normative influences, energy attitudes, energy knowledge, household behavior, lifestyle, and, of course, dwelling characteristics (heating system, insulation, sun orientation) and climatological factors (outside temperature). The purpose of this chapter is to analyze which factors influence consumer energy consumption. Given the nature of this study the emphasis will be on primarily nontechnical factors. As such this chapter is an attempt to answer research problem one.
Empirical psychological and sociological energy studies will be consulted to determine the actual strength of the relationships concerned. (1) If meaningful, these outcomes will be related to results from behavioral studies on correlates and determinants of proenvironmental behavior in general.

First of all, however, a number of behavioral science theories will be discussed which can clarify the micromotives behind energy consumption and their - often unintended - macroconsequences (Section 2.2). These theories describe "overconsumption" of energy as a result from a fundamental conflict between short-term individual benefits of energy consumption and long-term social or collective benefits of restricted consumption. In other words, some hypotheses are offered which explain why rational, self-interested consumers will not voluntarily restrict their consumption of energy solely because it is socially desirable.

2.2 Energy consumption as a commons dilemma: micromotives and macroconsequences

2.2.1 Social traps

Behavioral scientists have repeatedly pointed out the phenomenon that many societal problems, such as overpopulation, air pollution from automobiles, crowding in national parks, depletion of nonrenewable resources, overfishing, are characterized by a structural similarity (cf. Brechner & Linder, 1981; Ester & Leeuw, 1978; Hardin, 1968; Hardin & Baden, 1977; Olson, 1965; Platt, 1973; van den Doel, 1979; Wippler, 1977). This similarity is shaped by an underlying basic conflict of weighting short-term individual costs and benefits against long-term social costs and benefits.
More precisely, this structural similarity is that if confronted with this basic conflict, people tend to choose in favor of short-term individual benefits, a choice which if made collectively generates serious long-term social costs. Thus, environmental damage in national parks due to overcrowding of visitors is a consequence of the collective preference for short-term individual recreational benefits over long-term environmental costs. Air pollution caused by extensive use of private cars is an unintended consequence of the collective preference for individual transportation means.
Such individual cost-benefit decisions result in what Platt (1973) has called social traps. In general, social traps refer to "situations in society that contain traps formally like a fish trap, where men or oganizations or whole societies get themselves started in some direction or some set of relationships that later prove to be unpleasant or lethal and that they see no easy way to back out or to avoid... The trap is that each individual... continues to do something for his individual advantage that collectively is damaging to the group as a whole" (Platt, 1973, p. 641). Countertraps, on the other hand, are situations in which individual benefits block group-oriented behavior.
Applying Skinnerian reinforcement theory Platt uses a simple S (stimulus) - B (behavior) - R (reinforcement) sequence to formalize social traps and countertraps:

Trap: $S^{B}R_{s}+ \,..\, R_{l}-$

Countertrap: $S^{B}R_{s}- \,..\, R_{l}+$

where: R_s+ = short-term rewards
R_s- = short-term punishments
R_l+ = long-term rewards
R_l- = long-term punishments

Thus, both social traps and countertraps occur when there is a conflict between immediate individual consequences of behavior and long run group consequences. The difference between the two stems from the fact that a social trap results from an opposition between short-term positive individual consequences and long-term negative group consequences ($S^B R_s + > R_l -$), whereas a countertrap arises from an opposition of short-term negative individual consequences and long-term positive group consequences ($S^B R_s - > R_l +$).
In other words, the social trap phenomenon describes why people behave in ways which are collectively disadvantageous, and the countertrap phenomenon why they do not change their behavior in order to avoid or eliminate social trap situations. (2)
It is important to emphasize in this context that a social trap situation is not necessarily a simple aggregation of individual effects. Often complex transformation rules are needed to explain collective (sociological) phenomena from individual (psychological) phenomena (cf. Lindenberg, 1976; van Raaij, 1982; Schelling, 1978).

Applying Platt's analysis of social traps to the energy question, one might reason that from a behavioral point of view "overconsumption" of energy is a macroconsequence of a social energy trap. In proper terminology: there is an imbalance between short-term individual benefits from unresticted energy consumption and long-term social costs of such consumption ($S^B R_s + > R_l -$).
The following example given by Liebrand (1982) may illustrate this phenomenon. "In a small village of 150 inhabitants in the north of the Netherlands, an energy deficiency situation arose because of the severe winter of 1978/1979. The village was cut off from the power supply by heavy snowfall. However, the local blacksmith owned a generator, which permitted a short-term technical solution: everybody could use one but only one lamp and could operate a central heating pump, if they had one. The village's two farmers could also alternate the milking of their cows (since electrical equipment was used). However, this solution to the 'energy crisis' did not turn out to be a viable one -- so many villagers violated the preceding conditions that the generator was overloaded. In order to be able to continue the electricity supply, one villager regularly had to go out and identify those who violated the conditions agreed upon, a situation lasting until the connection to the power supply was repaired" (Liebrand, 1982, p.1).
This real life example shows that obviously it was the villagers' individual

interest to have more than one lamp, whereas their common interest dictates them to restrict themselves to one lamp. As it turned out, individual interests overruled common interest. The outcome, however, is that pursuing the individual interest leads to a disastrous situation in the long run not only from a common interest perspective but also from an individual interest perspective. Platt identifies three major subclasses of traps:

a. the one-person trap: a self-trap resulting from a time delay between the occurrence of certain behaviors (e.g. smoking, eating) and its long run consequences (e.g. lung cancer, overweight).

b. the missing hero-trap: only one person is needed to eliminate a collective bad but nobody is willing to do so (the bystander effect).

c. the Commons type trap: the common pursuit of individual benefits results in collective bads (e.g. depletion of natural resources, the Prisoner's Dilemma).

The Commons type trap especially seems to be relevant for understanding why individuals engage in activities which are collectively damaging, or, applied to our subject, why they favor unrestricted energy use - leading to the collective exploitation of a finite, nonrenewable energy resource like fossil energy - over restraint which would be socially desirable.
Hardin, in his by now classic essay "The tragedy of the commons" (1968), compared this situation to that of a group of shepherds grazing their herds on a common pasture. Each shepherd is motivated to increase the size of his herd, since the net profit of adding one animal will be entirely his. On the other hand, however, the negative consequences in terms of additional grazing created by one more animal is shared by all shepherds given the common property rights of the pasture. Therefore, taking into account the individualized positive consequences and the socialized negative consequences, a rational herdsman will conclude to increase his herd.
As this conclusion is reached by each herdsman, the logic of the commons dictates each herdsman to expand his herd which, as a consequence leads to overgrazing in the long run and, ultimately, to the ruin of both the pasture and shepherds themselves. "Therein is the tragedy. Each man is locked into a system that compels him to increase his herd without limit - in a world that is limited. Ruin is the destination towards which all men rush, each persuing his own best interest in a society that believes in the freedom of the commons. Freedom in a commons brings ruin to all" (Hardin, 1968, p. 138).

Hardin believes that many societal problems appear to have formal similarities to the commons trap, e.g. overpopulation, environmental deterioration, exploitation of scarce raw materials (cf. Hardin, 1968; Hardin & Baden, 1977). Traditional solutions (e.g. moral appeals, responsibility prompts) are ineffective according to Hardin to control commons problems. He strongly believes that only a special form of coercion, i.e. coercion mutually agreed upon by the majority of the people affected, can prevent them to act in accordance with the tragedy of the commons.

Though many theoretical and ethical arguments have been raised against Hardin's reflections on and solutions to commons problems (cf. Hardin & Baden, 1977; Edney, 1980), only a limited number of studies have empirically tested their underlying assumptions and predicted outcomes. "Though the hypothesis of the tragedy of the commons has an immediate intuitive appeal, which the recent concern for the environment serves to dramatize, there is no strong evidence that we know of, other than that supplied by hindsight, which tends to support the notion" (Wilson, 1977, p. 96). One of the most interesting exceptions is a study by Acheson (1975) on commons situations in the Maine lobster industry. (3) It seems that due to commons problems lobster landings in the New England area have decreased very rapidly in recent years. However, the decrease appears to be highly differential: some fishermen are better off than others. The reason for this is that the fishermen developed informal territorial rules. In some parts along the central Maine coast these territorial arrangements are highly nucleated, whereas in other parts one finds perimeter-defended areas. In <u>nucleated</u> areas there exist informal <u>common</u> fishing rights to the waters which become substantially weaker the further one goes from the harbour. Thus, fishing rights in these areas are commonly shared but in an uncontrolled manner. <u>Perimeter-defended</u> fishing areas are characterized by the fact that fishing rights are <u>privatized</u>, i.e. informal individual ownership rules exists. These individual fishing rights are traditionally held and inherited patrilineally, as are land property rights. The critical difference between these two non-legalized informal property rights systems is, that <u>entry</u> to perimeter-defended fishing areas is highly controlled, whereas the opposite is true for nucleated fishing areas. As such nucleated areas formally resemble commons situations. In order to see whether fishermen from nucleated areas behaved according to the Tragedy of the Commons, Acheson compared in his study fishing efforts from lobstermen in both areas. His findings are meaningful. Using both biological and economic parameters it appears that lobsters caught in the controlled perimeter-defended areas are larger than those caught in uncontrolled nucleated areas, also more lobsters are caught per trap, and reproduction possibilities are better because of higher stock density. Finally, there is a large differen-

ce in income: lobstermen from controlled fishing areas have higher gross income (up to 30%) than lobstermen from uncontrolled fishing areas.
Thus, common property rights seemed to produce less biological and economic benefits. (4) From a theoretical point of view the interesting aspect of Acheson's study is that dividing a common resource pool into individual pools redefines the problem of common resource control into one of self-control. In a recent laboratory study by Cass and Edney (1978) it was found that individualizing resource territories - in combination with increasing resource visibility - is an effective strategy for avoiding resource mismanagement.

2.2.2 Prisoner's Dilemma

The study by Acheson is still one of the few field experimental tests of the commons dilemma in a natural setting (Brechner & Linder, 1981). There is, however, a substantial amount of laboratory research on a special sub-class of commons problems, namely the Prisoners's Dilemma (cf. Dawes & Mactavish, 1975; Kelley & Grzelak, 1972; Luce & Raiffa, 1957). (5) In general, a Prisoner's Dilemma refers to a situation in which actors are isolated from each other, a situation which prevents interactor communication about behavioral choices. The payoff matrix for these choices is designed in such a way that when actors choose behavioral alternatives which are best in terms of their individual interest, their decision produces outcomes which are collectively damaging. The fact that in such situations interactor communication is not possible, implies that a mutual action strategy aimed at maximizing collective benefits cannot be developed. In other words, the crucial aspect seems to be that in a situation in which people cannot influence each other's decisions, individual rationality does not lead to collective rationality (van den Doel, 1978, p. 76).
It is exactly this last point which clarifies the anology between Prisoner's Dilemmas and individual cognitive decision processes with regards to energy conservation: as consumers cannot directly influence other consumers' decisions with respect to diminishing their consumption of energy, they are not willing to drastically conserve on their own consumption (cf. Johnston, Cooper & Page, Jr., 1981; Leeuw & Ester, 1981; Liebrand, 1981, 1982; Punter, 1980; Visser, 1980). Heertje (1977, p. 44) calls this the "After-You effect".
In their review of survey data and experimental studies on consumer energy conservation and commons dilemmas, Johnston, Cooper and Page, Jr. (1981) found empirical support for the existence of this effect (see also Harris, 1979). "The consumer is concerned that if he sacrifices to conserve, others will enjoy lower cost and available energy without having to conserve, a basic tenet of

the tragedy of the commons". (Johnston, Cooper & Page, Jr., 1981, p. 9).
A practical example of the fact that these perceptions consumers have of each other may be an important social barrier to energy conservation is provided by energy consumption in so-called master-metered apartments. In master-metered apartments consumers pay a fixed rent that includes the cost of utilities. As consumers have no direct financial incentive to conserve and as they expect other consumers not to conserve either, energy consumption in master-metered apartments is considerably higher than in individually-metered dwellings. Estimates indicate that master-metered dwellings use 20-50% more electricity than otherwise similar individually-metered apartments (cf. Cook, 1978; McClelland & Canter, 1979; McClelland & Cook, 1981b).
As far as energy and energy-related topics are concerned a number of the above-mentioned laboratory studies used research designs formally like social traps or Prisoner's Dilemmas. Foddy (1974) simulated Hardin's tragedy of the commons example (individual decision making of shepherds grazing their herd on a common pasture) in a laboratory setting. He was especially interested in the effects of public disclosure of individual choices and of sanctioning selfish choices. It turned out that these independent variables did not inhibit competitive behavior. In a similar study Brechner (1976) hypothesized that pool size and intragroup communication influences resource regulation. Effects were found but not in the predicted direction: resource depletion was greatest when the resource pool was small and with no communication allowed. The initial hypothesis of an interaction effect between pool size and communication - in the sense that resource regulation would be most effective within a small pool with communication possibilities - had to be rejected.
Van Run and Wolters (1980) observed in their experimental study that in a simulated commons situation intragroup communication increases conservation efforts (cf. Liebrand, 1981, 1982; Punters, 1980). Jerdee and Rosen (1974) and Edney and Harper (1978b) found similar results. As such their findings are more in line with more general studies showing that communication facilitates cooperative behavior in a commons situation or Prisoner's Dilemma (e.g. Dawes, McTavish & Shaklee, 1977; Grzelak & Tyszka, 1974; Wrightsman, O'Conner & Baker, 1972).
Stern (1976) observed in a laboratory commons game modeled on a carpool that pricing produced conservation efforts. Rationing and direct payoffs turned out to be ineffective. Interestingly, a mere educational strategy providing subjects with information about the long-term consequences of their actions also increased conservation. Recent experimental studies, however, indicate the striking persistence of group-damaging, self-interested individual behavior despite information about possible group consequences of such behavior (e.g. Cass & Edney, 1978; Edney & Harper, 1978c).

The role of information and education with respect to energy conservation will be discussed in following chapters.

2.2.3 The logic of collective action

From a history of science point of view it is interesting to observe that comparable theories on individual behavior and social consequences have been developed in a wide range of behavioral sciences, including psychology, sociology, human ecology, and political economy. An example of such a theory developed in political economy is Olson's (1965) collective action theory. The reason this theory will be discussed here at some length is that a growing number of behavioral scientists applied this theory to environmental issues in general and energy conservation in particular (cf. Edney, 1980; Ester, 1977, 1979a, 1979b; Ester & Leeuw, 1978; Ester & van der Meer, 1979a, 1982; Leeuw & Ester, 1981; van der Meer, 1981; Visser, 1980). In addition to this particular application the theory has proven its strength in explaining a number of policy frictions in the collective sector (cf. Buchanan, 1968; Ester & Leeuw, 1978; Frohlich & Oppenheimer, 1978; Leeuw, 1981; Lulofs, 1978a, 1978b; Ultee, 1977; van den Doel, 1980).

Essential in Olson's theory is the distinction between private goods and collective goods. Private goods are goods with property rights such that other consumers can be excluded from consumption. Furthermore, private goods can be divided up among consumers. As such, excludability and divisibility are the defining characteristics of private goods (cf. Head, 1962). Collective goods, however, are nonexcludable and technically or economically nondivisible. Olson defines a collective good as "...any such good such that, if any person $x_1,\ldots,x_i,\ldots,x_n$ consumes it, it cannot feasibly be withheld from the others in that group" (Olson, 1965, p. 14). Examples of collective goods are: national defense, nonpolluted air, police protection.

Olson's main argument is that although the majority of the population has a common interest in the provision of collective goods, nobody is voluntarily willing to contribute. This is precisely so because of the binding characteristics of collective goods: non-contribution does not lead to exclusion. Because consumers cannot be excluded from collective goods and because consumption of collective goods is independent of one's own contribution to the provision of these goods, it is irrational - according to Olson - for an individual consumer to contribute voluntarily. Even if one does not participate one cannot be excluded from the benefits of the collective good.

The problem of so-called "free rider behavior" is therefore a crucial characteristic of the friction between individual and collective action (cf. Buchanan, 1968, p. 86-92; Frohlich, Oppenheimer & Young, 1971, p. 12-18; van den Doel, 1975, p. 63-65). Such behavior is characterized by the fact that everyone is in favor of the provision of collective goods but no one has an individual interest in paying the costs.

It can be predicted that the larger the group the more such behavior will occur: the expectation that <u>other</u> group members will participate in the provision of collective goods varies with group size. Moreover, perceived efectiveness of one's own contribution will be smaller in large groups since "...the larger the group, the less the likelihood that the contribution of anyone will be perceptible" (Olson, op. cit., p. 45). (6) Olson's alternative hypothesis states that "...unless the number of individuals in a group is quite small, or unless there is coercion or some other special device to make individuals act in their common interest, <u>rational, self-interested individuals will not act to achieve their common or group interests</u>" (Olson, op. cit., p.2).

The taxation example may serve to illustrate this hypothesis. Verbally most people are not against paying taxes. Nevertheless, paying taxes is <u>compulsory</u> because of the plausible assumption that no rational citizen is willing to pay in case of voluntary taxation. Thus, although tax money is used for the supply of collective goods, nobody is willing to make financial contributions on a voluntary basis.

In order to explain why people under certain circumstances do in fact participate in the provision of collective goods, Olson developed his "by-product and special interest theory" (Olson, 1965, chap. 6). This theory states that large groups can provide their group members with <u>selective stimuli</u>, i.e. noncollective, private goods compensating the individual costs of participation. Characteristic for these stimuli is that their receipt depends on whether or not the individual group member participated in the supply of the collective good. In other words, an incentive system is designed which reinforces individual participation. Examples of positive selective stimuli are: patronage dividends, status-enhancing jobs in trade unions and political parties, or tax incentives for adopting energy conservation innovations. These stimuli may also be of a punishing nature: compulsory tax paying, forced trade union membership ("closed shop"), or legal sanctions for noncompliance with mandatory conservation requirements.

Apart from the intuitive appeal of Olson's logic of collective action theory, very few studies have empirically tested its validity. In a simulation study of coalition processes by Burgess and Robinson (1969) and in a laboratory experiment by Chamberlin (1978) strong empirical support was found for this theory, but a series of experiments by Marwell and Ames (1979, 1980) yielded only weak

support. (7) Recently, attempts have been made to reconstruct Olson's theory in terms of informative hypotheses (Ester & Leeuw, 1978). In order to show the relevance of this theory for our subject, these hypotheses will be reformulated and applied to energy conservation (cf. Leeuw & Ester, 1981).
If energy conservation can indeed be defined as a contribution to the provision of a collective good (i.e. reducing resource depletion) than the logic of collective action theory is able to generate some conditions under which consumers are and are not willing to conserve energy (cf. Visser, 1980).
First of all, two general predictions can be made:

1. consumers are not willing to conserve energy solely on a voluntary basis.
2. the provision of selective stimuli is necessary to promote consumer energy conservation.

At least the following six hypotheses can be derived from Olson's theory with respect to energy conservation in the consumer sector (cf. Leeuw & Ester, 1981, p. 52-53). In each of these hypotheses it is assumed that subjects have a common interest in energy conservation as a means to reduce fossil energy depletion. The reason this phrase is left out in each hypothesis is simply a matter of elegance.

H_1: if in a large group conservation efforts are not attended with positive or negative selective stimuli, group members will not optimally conserve energy.

H_2: in a small group, however, group members will conserve more energy as compared to the situation under H_1.

These hypotheses can be explained through the following theory:

H_3: the larger the group, the smaller the contribution group members expect to be able to make to collective energy conservation efforts and the more they expect other members to take conservation efforts.

H_4: the smaller the contribution members of a large group expect to be able to make to collective conservations efforts and the more they expect other members to take conservation efforts, the less optimal these members will conserve energy.

H_5: the larger the group, the less visible group members perceive their own contribution to collective energy conservation efforts.

H_6: the less visible members of a large group perceive their own contribution to collective energy conservation efforts, the less optimal these members will conserve energy.

2.2.4 Solutions to social traps

This chapter started by applying Platt's theory on social traps to energy consumption, now the circle will be closed in ending this section by outlining his solutions to social traps. Platt (1973, p. 648-650) suggests six general solutions or strategies to prevent or get out of social traps.

1. change of delay of consequences: this solution essentially converts long-term consequences of individual behavioral choices into more immediate ones. A concrete application of this principle to social energy traps would be to give individual consumers frequent feedback on their energy consumption in order to make consequences more immediate and salient. One might also think of educational strategies which could make the long-term macroconsequences of unrestricted energy consumption more visible. (8)

2. add counterreinforcers: (punishments or rewards) to encourage or discourage certain behaviors. Energy examples of this strategy are mandatory residential insulations norms, energy surtaxes, financial incentives for adopting energy conserving technologies (e.g. tax deductions for solar energy boilers, low-interest loans for home insulation).

3. change the nature of long-run consequences: this appears to be a more technical or structural strategy in the sense that situational improvements such as design or planning alterations might change the nature of long-term consequences. Examples are: home insulation, increasing energy-efficiency of household appliances, household energy monitors.

4. add short-term rewards for competing behavior: this solution suggests to reinforce alternative behaviors which avoid the negative long-term consequences of initial behaviors. One might think of immediate financial reinforcement of energy conservation efforts through e.g. cash bonuses or cash rebates.

5. get outside help in changing the reinforcement patterns of locked-in loops: in this strategy skilled outsiders offer assistance to change the locked-in reinforcer network. (9) Possible energy examples of this strategy are: residential energy audits for consumers by energy specialists, consulting "significant others" about energy conservation techniques.

6. set up a superordinate authority to regulate behavior to prevent or solve social traps situations. Characteristic of this strategy is that a higher level

of decision-making is called into existence "to present entrapments, to allocate resources, to mediate conflicts, and to redirect immediate reinforcement patterns to more rewarding long-range goals" (Platt, 197, p. 650). Examples are: arrangements between oil-importing countries on maximum oil imports or on common conservation target goals. Forced energy rationing is another example.

Although these strategies or solutions are not exhaustive (note that Platt does not mention territorization or privatization of resources) and have been critisized by several authors, (10) there is nevertheless a close relationship between the logic of social traps as analyzed by Platt and the logic of solutions he suggests. Also, his analysis permits the formulation and deduction of quite specific policy recommendations.
In Chapter 3 when discussing behavioral interventions for promoting residential energy conservation, we will return to Platt's analysis of social traps.

Summary
In this section a number of general theoretical perspectives has been outlined explaining conflicts between individual and collective rationality applied to the issue of "overconsumption" of energy. Attention has been paid to Platt's theory of social traps, theories of commons dilemmas and Prisoner's dilemmas, as well as to Olson's theory of the logic of collective action. If available, empirical evidence has been presented to establish the scientific status of the theory and to indicate its usefulness for understanding micromotives and macroconsequences of energy consumption. Also strategies to prevent or solve social trap situations were discussed.
The rationale behind the rather general and abstract character of the discussion so far is to lay down a theoretical framework on which in the next section a more specific analysis of correlates of consumer energy consumption can be raised.

2.3 Personal correlates of energy conserving behavior

In the remaining part of this chapter some nontechnical personal correlates of energy conserving behavior will be discussed: general energy attitudes and value orientations (Section 2.3.1), specific energy conservation attitudes (Section 2.3.2), energy knowledge (Section 2.3.3), and consumers' lifestyle and sociodemographic characteristics (Section 2.3.4). Insight into the empirical strength of these correlates is important since they may be crucial elements of

consumers' responsiveness to behavioral interventions aimed at promoting residential energy conservation.
Findings on these personal correlates will be briefly related to outcomes of studies on correlates of environmentally-conscious consumer behavior in general.

2.3.1 General energy attitudes and value orientations

Since the 1973 energy crisis, numerous survey studies have been conducted by social scientists to explore people's general attitudes toward several dimensions of the energy issue, including belief in the existence of an energy crisis, perceived saliency of energy scarcity, perceived personal and social impacts of the energy question, attitudes toward energy conservation, attitudes toward energy policy options, and values related to energy attitudes (e.g. Anderson & Lipsey, 1977; Bartell, 1974; Brunner & Bennett, 1977; Bultena, 1976; Cunningham & Lopreato, 1977; Durand, 1979; Gottlieb, 1978; Morrison, Keith & Zuiches, 1978; Neuman, 1982; Perlman & Warren, 1977; Philips, 1976; Ritsema, Midden & van der Heijden, 1982; Thompson & Mactavish, 1976; Zuiches, 1976). As these studies have been reviewed elsewhere (Cunningham & Lopreato, 1977; Milstein, 1976; Neuman, 1980; Olsen, 1981; Olsen & Goodnight, 1977; Warkov, 1978), this section will restrict itself to drawing some major conclusions.

According to U.S. surveys, between 40 and 60% of consumers believe that the energy crisis is a serious, credible and long-term problem, whereas 20 to 40% is somewhat sceptical in this respect (Olsen, 1981). A recent survey in the Netherlands (Ritsema, Midden & van der Heijden, 1982) indicates that two-third of the Dutch population is convinced of the reality of the energy problem, though about one-fourth is not concerned at all. Sixty-two percent of the respondents expects a serious energy shortage in the Netherlands within 30 years, and 40% within 10 years. Possible causes of energy shortages are primarily related by respondents to political factors (energy politics by Arab countries and national governments, oil companies policy) and in a lesser degree to technological factors. Energy policy options favored by respondents include solar and wind energy (89%), energy conservation (86%), coal (70%), whereas only 28% is in favor of expanding the use of nuclear energy (60% is opposed). It was demonstrated that respondents hold very positive attitudes toward energy conservation in the household, though the underlying psychological mechanism seems to be that <u>other</u> consumers should take the lead: 73% agrees with the statement that households have to conserve more energy, whereas at the same time 83% indicates that they themselves conserved enough energy. This psychological

mechanism has some resemblance with the phenomenon of the "After-You" effect (Heertje, 1977) which was mentioned in the first part of this chapter, and has been demonstrated in many studies.
Most surveys show that the single most important motive for residential energy conservation is price (e.g. Cunningham & Lopreato, 1977; Gottlieb & Matre, 1976; Perlman & Warren, 1977), though social motives may play a role too (Olsen & Cluett, 1979; Ritsema, Midden & van der Heijden, 1982), provided that these social motives are internalized by consumers.
Surveys have demonstrated, in general, that consumers are willing to make efforts to conserve energy in the household as long as they are not forced to spend substantial sums of money or perceive negative impacts on their personal lifestyle (Cunningham & Lopreato, 1977). This means that energy conservation intentions are influenced or guided by cost/benefit considerations.
As far as attitudes toward governmental energy conservation policies are concerned, many studies have discovered a similar pattern: consumers prefer voluntary policies over mandatory policies, and incentives over penalties (cf. Bennett & Moore, 1981; Milstein, 1978).
According to Olsen (1981) energy conservation policy acceptance is influenced by a number of other considerations: awareness of the overall ecological situation, acceptance of personal responsibility and/or expectations of personal consequences in regard to the energy crisis, favorable attitudes toward the current administration, and equity considerations.(11)
The relationship between human values and conservation attitudes is a neglected area in social science energy research (cf. Neuman, 1980, 1982). This has to be regretted, since many observers have related the energy problem to current value systems, in the sense that there seems to be a basic conflict between these value systems and the need of a conservation ethic (Illich, 1973; Schumacher, 1973). Nevertheless, some authors have observed an emerging set of conservation values in modern society, values, for example which express the need of lifestyles based on "voluntary simplicity" (Elgin, 1981) characterized by material simplicity, self-determination, ecological awareness, human scale, and personal growth (Elgin & Mitchell, 1977). Based on recent national U.S. polls, Olsen (1981) found some evidence for this emerging set of conservation values, containing the following elements: less emphasis on material possessions, rejection of economic growth as a national goal, rejection of technology as the only way of solving contemporary problems, environmental protection, and more attention to quality of human life.

The basic question is of course whether these general energy attitudes and value orientations are related to conservation behavior. The existing evidence

is rather conclusive: most studies do not report significant relationships between general energy attitudes and conservation behavior (Anderson & Lipsey, 1977; Bartell, 1974; Cunningham & Lopreato, 1977; Gottlieb, 1978; Heslop, Moran & Cousineau, 1981; Morrison, Keith & Zuiches, 1978; Nietzel & Winett, 1977; Lopreato & Meriweather, 1976; Perlman & Warren, 1977; Sears, Tyler, Citin & Kinder, 1978; Thompson & MacTavish, 1976; Verhallen & van Raaij, 1981). Given the limited number of studies on the influence of human values on conservation behavior, no firm conclusion is possible. Neuman (1982) found that personal values (modified version of Rokeach's value survey) were not or only weakly associated with conservation efforts. Leonard-Barton (1981), however, observed that a voluntary simplicity lifestyle was positively related to investment in energy-conserving equipment (e.g. wall insulation) and to intentions to purchase residential solar equipment.
Apart from the explanation that will be offered in the following subsection, one might suggest that the saliency of such general attitudes is not strong enough to generate energy behaviors which are congruent with these attitudes. Also, situational constraints are typically overlooked in the surveys mentioned.

Findings from studies on the relationship between environmental attitudes and environmental behaviors have also provided evidence for low or even zero correlations between general environmentally-concerned attitudes and specific ecologically-consious behaviors (cf. Ester, 1979b; Ester, 1981; Lipsey, 1977; O'Riordan, 1976).

2.3.2 Specific energy conservation attitudes

Recent developments in attitude-behavior theory and research have convincingly demonstrated that attitudes are relatively good predictors of behavior only when the attitudinal and behavioral measures show a high degree of correspondence (cf. Ajzen & Fishbein, 1980; Cialdini, Petty & Cacioppo, 1981; Eagly & Himmelfarb, 1978; Fishbein & Ajzen, 1980; Hill, 1980). This means that attitudes and related behaviors have to be measured at an equivalent level of specificity with respect to action, target, context, and time dimensions. This theoretical notion can very well explain why in the preceding subsection it was found that general consumer attitudes towards energy issues are not related to specific energy conserving behaviors: there is no corresponding level of specificity between attitudes and behaviors. The hypothesis can then be formulated that corresponding levels of attitudinal and behavioral measurements will yield higher correlations between energy conservation attitudes and energy conserving behaviors (see also Chapter 4). In this subsection results of some studies will be mentioned to explore the degree of empirical support for this hypothesis.

In two studies by Seligman, Darley and Becker (1978) and Seligman, Kriss, Darley, Fazio, Becker and Pryor (1979), a rather successful attempt has been made to explain residential electricity consumption by homeowners' attitudes toward electricity use. In a first study among a sample of 56 couples living in a planned unit development in New Jersey a factor analysis of 28 attitude questions with respect to air conditioning usage (target behavior) revealed four factors: attitude toward personal comfort and health, high effort - low pay off (limited dollar savings from conservation efforts), attitude toward the role of the individual to alleviate the energy crisis, and concern with the legitimacy of the energy crisis. An overall multiple regression analysis indicated that 55% of the variance in electricity consumption was accounted for by these four factors. The contribution of the first two factors turned out to be the most important ones (R^2 = .30 and .25, respectively). A second survey of 69 couples in the same community confirmed the results of the first study. A factor analysis revealed the same factors emerged plus an additional factor: "belief in science and technology". An overall multiple regression analysis showed that 59% of the variance in electricity use could be explained by these attitudinal predictors, However, in this second study the personal comfort and health factors were the only statistically significant predictors (R^2 = .42 and .11, respectively). It must be noted that other studies using specific attitudinal predictors have been somewhat less successful in explaining residential energy use.

A study by Midden, Weenig, Houwen, Meter, Westerterp & Zieverink (1982) revealed two important specific attitudinal factors: the energy question as a political problem (both in the sense of responsibility attribution and perceived personal effectiveness) and concern with energy scarcity. It was found that attitudinal variables accounted for 18% of the variance in natural gas consumption and 24% of the variance in electricity consumption (See also Section 3.3.1.1).

In another study by Ritsema, Midden and van der Heijden (1982) among a national sample from the Dutch adult population four major factors were observed in specific belief structures with respect to residential energy conservation: negative consequences of energy conservation, positive consequences of energy conservation, responsibility acceptance, and concern with clean and safe energy production. Using the Ajzen and Fishbein model (Ajzen & Fishbein, 1980; Fishbein & Ajzen, 1975) it was found that attitudes and social norms with respect to energy conservation explained 24% of the variance in conservation intentions, which in turn accounted for 61% of the variance in energy conserving behaviors.

A study by Kok, Abrahamse, Douma, Langejan, Sietsma, Slob and de Vries (1979) among a random sample of the Dutch population, using similar attitude measurements as the Ritsema et al. (1982) study, indicated that specific attitudes and social norms with respect to energy conservation explained 24% of the variance in conservation intentions and 20% of the variance in conservation behaviors (see also Kok, 1981).

Thus, there is some empirical evidence for the hypothesis that specific energy conservation attitudes are better predictors of energy conserving behaviors than general attitudes toward or beliefs about energy issues. Studies in other ecologically-relevant behaviors tend to support the necessity of corresponding levels of attitudinal and behavioral measures (cf. Ester & van der Meer, 1979b; 1982; Leonard-Barton & Rogers, 1979; van der Meer, 1981; van der Meer & Berghuis, 1976; Weigel, Vernon & Tognacci, 1974).

2.3.3 Energy knowledge

A central hypothesis underlying many governmental and nongovernmental educational programs aimed at promoting household energy conservation seems to be, that improving consumers' knowledge of residential energy matters and conservation practices will increase conservation efforts by consumers. This hypothesis is based on two assumptions: a) consumers are generally ill-informed about energy matters and conservation practices, and b) knowledge of residential energy matters and conservation practices is positively related to conservation efforts. This section will try to establish the empirical validity of both assumptions.

Most studies on consumers' energy knowledge tend to support the assumption that in general there is a clear lack of accurate knowledge (Cunningham & Lopreato, 1977; Farhar-Pilgrim & Shoemaker, 1981; Neuman, 1980; Olsen & Goodnight, 1977). Ellis and Gaskell (1978) have introduced the concept of "energy illiteracy" to describe this lack of knowledge. In spite of many educational programs, the situation seems to be that consumers are rather ill-informed about residential energy issues and conservation possibilities, though there is some empirical evidence which shows improved knowledge levels after the 1973 energy crisis (Olsen & Goodnight, 1977).

Milstein (1976, 1978) found in a national U.S. survey on energy issues that 36% of the respondents did not know that lower-wattage light bulbs use less electricity, 59% incorrectly thought that keeping a light bulb on uses less elec-

tricity than turning it off several times an hour, and 42% of respondents with water heater controls did not know its setting (see also Kilkeary, 1975; Rappeport & Labaw, 1974). In a survey by the National Center for Education Statistics (1978) among a national probability sample of young adults in the United States it was found that although young adults demonstrated quite a bit of concern with the energy question, they showed little understanding and knowledge of energy-related matters.
In a recent national survey in the Netherlands by Ritsema, Midden and van der Heijden (1982) it was found that only 20% of the respondents knew the correct price of one cubic meter natural gas, and that people tend to underestimate cost of warm water and the effects of thermostat set-backs, whereas they tend to overestimate cost of lighting. Results of another Dutch study conducted in the area of the Hague (Geradts & Geradts, 1978) show that about 75% of the sample either did not know the correct natural gas and electricity price or gave incorrect answers. Two other Dutch studies (van Amstel, Ester, van Schijndel & Schreurs, 1980; Kok, Abrahamse, Douma, Langejan, Sietsma, Slob & de Vries, 1979) also provide some evidence for the phenomenon of consumer energy illiteracy.

Although accurate energy knowledge may be a positive stimulus to consumers who are already engaged in proconservation behavior, there is no convincing empirical support for the assumption that energy knowledge has a direct influence on energy conserving behavior (Farhar, Weir, Unseld & Burns, 1979; Kilkeary, 1975; Kok, Abrahamse, Douma, Langejan, Sietsma, Slob & de Vries, 1979; Neuman, 1980; Ritsema, Midden & van der Heijden, 1982). "It seems doubtful that knowledge alone constitutes a powerful determinant of resource conserving behavior. Specific information on how to conserve likely represents a necessary but not sufficient condition to do so ..." (Neuman, 1980, p. 13). At best, energy knowledge is a mediating factor between energy attitudes, conservation intentions and conservation behavior.

Studies on other ecologically-relevant behaviors lead to similar conclusions: consumers generally have low levels of environmental knowledge, and environmental knowledge is not directly related to environmental behavior (cf. Ester, 1979b; Ester & van der Meer, 1979b, 1982; Lipsey, 1977; Maloney & Ward, 1973; Maloney, Ward & Braught, 1975).

2.3.4 Lifestyle and sociodemographic characteristics

In Chapter 1 a number of studies was mentioned which all concluded on the basis of empirical evidence that energy-related lifestyles and behavioral routines of consumers are important factors in explaining differences in residential energy consumption (Sonderegger, 1978; Verhallen & van Raaij, 1979, 1980; Wotaki, 1977). It has been consistently observed that in similar dwellings, household energy use may differ by two or three fold (Geller, Winett & Everett, 1982; Winett, Neale, Williams, Yokley & Kauder, 1979). Van Raaij and Verhallen (1983) distinguish three types of energy-related behaviors in which these differences may occur: purchase-related behavior (e.g. purchase of consumer durables), usage-related behavior (e.g. thermostat setting, airing of rooms, air conditioning usage), and maintenance and operating behavior (e.g. heating system maintenance).

It has not only be found that different lifestyles are related to different levels of energy consumption (Mazur & Rosa, 1974), but also that consumers with different lifestyles react differently to changed external energy conditions. Dillman, Rosa and Dillman (1982), for instance, found that faced with rapidly increasing energy prices lower-income consumers tend to accept lifestyle cutbacks while higher income consumers invest in conservation actions (See also van Raaij & Eilander, 1983). This brings us to the relationship between sociodemographic variables and conservation efforts. The importance of sociodemographic variables in this respect is - among other things - related to the fact that they can be used as segmentation criteria for designing and implementing energy conservation programs. Three variables in particular will be looked at: income, education, and age.

Income

As noted before, price is the most important motive for consumers to conserve. Therefore, one would expect a relationship between income and conservation efforts. Evidence on the nature of this relationship, however, is quite diverse. A number of studies reported positive relationships between income and conservation efforts (Barnaby & Reizenstein, 1975; Bultena, 1976; Curtin, 1976; Milstein, 1976; Murray, Minor, Bradburn, Cotterman, Frankel & Pisarski, 1974; Olsen & Goodnight, 1977; Talarzyk & Omura, 1975; Warren & Clifford, 1975). Some studies, on the other hand, found an inverse relationship between income and conservation, with middle-income consumers demonstrating greatest conservation efforts (Kilkeary, 1975; Morrison, Keith & Zuiches, 1979; Walker & Draper, 1975). One plausible explanation with face validity for the existence of such a curvilinear relationship has been offered by Cunningham & Lopreato (1977). According to their explanation high-income consumers use more energy and can better afford to keep doing so at higher energy costs. At the same time they can decrease their energy consumption without major lifestyle implications. Low-income consumers, however, cannot do so since their energy consumption is already minimal and further cutbacks would create serious lifestyle threats. Middle-income consumers are therefore believed to be the most active income segment with respect to energy conservation, as they are very sensitive to price increases and are able to decrease their consumption of energy without major lifestyle alterations.

The plausability of this explanation depends of course on the correctness of the assumed positive relationship between income and energy consumption. A recent study in the Netherlands (Hemrica, 1981) using a random sample of the Dutch population, has seriously challenged this assumption, since there was no relationship between the two variables up to annual incomes of Dfl. 51,000.

Education

The relationship between education and conservation efforts turns out to be rather straightforward: most studies report positive relationships (Barnaby & Reizenstein, 1975; Bultena, 1976; Curtin, 1976; Gottlieb & Matre, 1976, Thompson & MacTavish, 1976; Zuiches, 1976). Although the correlations vary in strength, and are often related to different behaviors, as well as the fact that education covaries with income, the direction of the correlations is nevertheless obvious and consistent: higher education covaries with the likelihood of conservation efforts. A number of studies has also indicated that education is positively related to beliefs in the seriousness of the energy problem (Cunningham & Lopreato, 1977; Thompson & MacTavish, 1976).

Age

To some extent age interacts with income and education. Most studies, however, did not attempt to account for the separate effects of age with respect to energy conservation.

It appears that the relationship between age and conservation efforts is not consistent. Some studies have reported positive relationships (Cunningham & Lopreato, 1977; Kok, 1981; Morrison, Keith & Zuiches, 1979), whereas other studies found negative relationships (Curtin, 1975; Gottlieb & Matre, 1976). In addition, some studies have demonstrated a curvilinear relationship -i.e. with regard to some conservation behaviors - between age and conservation, though not necessarily with identical directions of the curve (Cunningham & Lopreato, 1977; Curtin, 1976; Lopreato & Meriweather, 1976; Talarzyk & Omura, 1975).

In summary, this section examined some personal correlates of consumer energy conservation behavior, including general energy attitudes, value orientations, specific energy conservation attitudes, energy knowledge, lifestyle and socio-demographic characteristics. It has to be emphasized that technical correlates, contextual constraints, as well as institutional barriers (cf. Crossley, 1982; Olsen & Joerges, 1981) - though obviously important determinants of consumer energy use - were not taken into account.
Generalizations from behavioral environmental research seem to reveal a somewhat similar pattern: proenvironmental attitudes and behavior are positively related to education, whereas the associations reported for income are again quite ambiguous and contradictory. It seems, however, that environmental research has more often shown positive correlations with age (Lipsey, 1977; van Liere & Dunlap, 1979).

Notes

1. This review is not meant to be exhaustive but rather presents a selective overview.
2. See Schelling (1971, 1977) for a captivating description of many situations in which short-term, self-interested behavior produces long-term collective disadvantages.
3. See also Wilson (1977).
4. See also Alchian and Demsetz (1973), and Furubotn and Pejovich (1974).
5. For the differences between social traps and Prisoner's Dilemmas see Brechner and Linder (1981, p. 42), and Stern (1976).
6. For a different view on the relationship between group size and supply of collective goods see Frohlich, Oppenheimer and Young (1971, p. 145-150).
7. See also Olson and Zeckhauser (1966).
8. Cf. Stern's experiment on a simulated car pool (Stern, 1976).
9. As far as this strategy is concerned Platt especially refers to Tharp and Wetzel (1969).
10. Edney (1980), for instance, argues that social traps and commons dilemmas are not just situations of conflicting rationalities, but primarily situations of conflicting value systems. He also blames Platt for having a rather mechanistic view on human behavior and for excluding "positive humanistic tendencies in people" (Edney, 1980, p. 134). Edney favors solutions to social traps and common dilemmas based on territory and trust. (See also Cass & Edney, 1978).
11. See also Olsen (1982).

3. BEHAVIORAL INTERVENTIONS FOR PROMOTING CONSUMER ENERGY CONSERVATION

3.1 Introduction

This chapter presents a detailed and critical review of behavioral experiments in energy conservation. Detailed because it serves to develop methodological directives for this study, and critical because a thorough analysis shows that both from a policy and research point of view many of the experiments to be reviewed leave much to be desired. This review and critical evaluation will be presented in Section 3.3.
First of all, however, some theoretical remarks will be made with respect to behavioral interventions aimed at promoting consumer energy conservation (Section 3.2). These interventions will be related to some of the theories outlined in the previous chapter on behavioral "causes" of energy problems.

3.2 Antecedent and consequence behavioral strategies: A general framework

In this section a number of behavioral strategies will be outlined which can be used for promoting energy conservation by consumers. As indicated in the previous section, these strategies will be related to the theoretical framework presented in Chapter 2.

First of all, a general distinction has to be made between physical technology and behavioral technology (Cone & Hayes, 1980) with respect to encouraging energy conservation. Physical technology refers to the development of primarily technical solutions, e.g. insulation measures, fuel efficient cars, high-efficiency boilers, whereas behavioral technology mainly focuses on behavioral solutions. Many observers have indicated that the present emphasis in most societies tends to be on the development of technological solutions and to a much lesser degree on the implementation of behavioral innovations (Cone & Hayes, 1980; Darley, 1978; Ester, 1979a; Geller, Winett & Everett, 1982). However, as put forward in Chapter 1, technical solutions certainly are necessary but not sufficient strategies for promoting energy conservation, since technical solutions have to be adopted by consumers in order to be effective.
This section will explore some of these behavioral solutions and strategies.

A distinction which is quite useful and fundamental in this respect is the one between antecedent strategies and consequence strategies (cf. Carlyle & Geller, 1979; Tuso & Geller, 1976). "Antecendent interventions can be defined as stimu-

lus events occurring before the target behavior, designed to increase or decrease the probability of the target behavior (...) Consequence interventions, on the other hand, can be described as stimulus events occurring after the target behavior, designed to increase or decrease the probability of the target behavior" (Ester & Winett, 1981, p. 202). Examples of antecedent strategies or interventions are: information to consumers about energy conservation and prompting consumers to conserve energy. Examples of consequence strategies or interventions are: feedback to consumers on their household energy consumption, material or immaterial rewards to consumers for decreased energy consumption, or punishments for increased consumption. Target behaviors may involve thermostat control, insulation measures, adoption of solar technology, purchase behavior of electrical household appliances, maintenance of heating systems, goal setting or public commitment, to mention just a few examples.
In figure 3.1 a number of behavioral technologies are listed which correspond with either antecedent or consequence strategies with respect to promoting energy conservation by consumers.

Figure 3.1: Major Antecedent and Consequence Strategies for Promoting Energy Conservation by Consumers

ANTECEDENT STRATEGIES	CONSEQUENCE STRATEGIES
- information/education e.g. mass media energy conservation campaigns	- feedback e.g. providing consumers with frequent information about changes in their energy consumption
- prompting e.g. tv spots prompting energy conservation	- self-monitoring e.g. teaching consumers how to monitor their own energy consumption
- modeling e.g. video modeling tapes about energy conservation	- reinforcement/punishment e.g. monetary incentives for energy conservation

It has to be added that the strategies mentioned in figure 3.1 do not represent an exclusive enumeration, they merely exemplify interventions that have been tested so far in behavioral energy research. In the next section we will carefully review behavioral experiments which evaluated these interventions. Also, one has to realize that a strict distinction between antecendent and consequence strategies is somewhat misleading (cf. Carlyle & Geller, 1979, p. 46). The response evoked by a consequence strategy may very well serve as an antecendent for a subsequent response. Self-monitoring, for example, can be defined as a consequence strategy which provides the consumer with information about energy consequences of certain behavior patterns, but this information may in turn function as an antecendent for subsequent behaviors.

3.2.1 Energy conservation information and prompts

One of the strategies that was suggested by Platt (See Chapter 2) to prevent or to escape from various social traps is to induce people to act from a long-term perspective by educating them about the consequences of their behavior. In the case of energy traps such an educational strategy is to provide consumers with information about several consequences of energy consumption (e.g. depletion of natural resources, environmental effects, social impacts), and to educate them how to conserve energy or to prompt energy-efficient behavior.

Undoubtedly, this information strategy is one of the most widely used energy conservation policy instruments by governments and utility companies in western countries (cf. Joerges a.o., 1982). Especially popular are mass media campaigns urging consumers to conserve energy. These campaigns may focus on raising consumers' general awareness of energy problems, the necessity of energy conservation or on prompting specific target behaviors (e.g. insulation, thermostat setting, ventilation behavior, efficient electricity use, energy-conscious driving behavior, purchase of high-efficiency boilers).

In Chapter 2 a number of studies were mentioned indicating that consumers generally are ill-informed about energy matters. Therefore, energy conservation information is certainly necessary if consumers are to increase their energy knowledge. Without adequate information and knowledge consumers can hardly be expected to practice proper energy conserving behaviors.

It is important to recognize that the effectiveness of informational interventions depends among other things on consumers' perception of the information and of the information source (credibility), their involvement in the information object, and the degree in which the information is tailored to their information needs (cf. van den Ban, 1982; van Woerkum, 1982).

The leading hypothesis underlying this informational strategy seems to be that informing consumers about energy conservation will change their attitudes toward energy conservation in a more favorable direction, which will in turn influence their energy behavior. However, empirical support for this hypothesis

tends to be rather weak since one of the conclusions from the preceding chapter was, that (general) energy attitudes are hardly predictive of energy behavior. Also, attitude change does not necessarily or logically precede behavior change. Bem (1970) has shown that in many cases attitude change follows behavior change. Also, one may wonder whether - in view of Olson's collective action theory - conservation information and prompts are perceived by consumers as powerful selective stimuli.
However, the popularity of informational strategies aimed at promoting energy conservation (e.g. through leaflets, television spots, newspaper prompts) is of course related to their potential large-scale applicability, the fact that they are relatively inexpensive, as well as that those strategies do not limit individual freedom and thereby hardly evoke public reactance.

3.2.2 Energy conservation modeling

An important aspect of socialization processes is that new behavioral patterns are acquired through observational learning. This means that behavioral patterns are learned through observing behaviors of others, who as such function as behavioral models. Applied to our subject, this principle implies that energy behavior is influenced by observing energy behaviors of others. Furthermore, observational learning theory seems to suggest that using models demonstrating energy conservation behavior could increase conservation practices by others.
Behavior modification research (Bandura, 1969) has shown that modeling plays a highly significant role in changing behavioral patterns. If new patterns are to be learned, actors must be provided with models demonstrating the desired behavior. It is important that actors are reinforced for showing new behaviors, and that actors held positive attitudes toward the models.
Though no detailed scenarios are available on how to utilize modeling for promoting energy conservation by consumers, at least one application seems interesting (cf. Geller, Winett & Everett, 1982, p. 161-164). Given the increasing importance in modern society of audio-visual media (television, video) for the dissemination of information and knowledge, one may very well use these media for developing programs on energy conservation based on modeling principles. These programs could focus on effective conservation strategies demonstrated by pre-selected models. In Section 3.3.2.1 two behavioral experiments will be described which provide empirical support for the potential effectiveness of videotape modeling in promoting residential energy conservation (Winett, Hatcher, Fort, Leckliter, Love, Riley & Fishback, 1982; Winett, Leckliter, Love, Chinn & Stahl, 1983). Unfortunately, these are - to our knowledge - the only two experiments conducted so far in this area.

3.2.3 Energy consumption feedback

Changing the delay of consequences of individual behavior by converting long-term consequences into more immediate ones, was one of the possible strategies offered by Platt to avoid social traps. A practical application of this strategy in the area of residential energy use would be to provide consumers with more frequent feedback on their energy consumption. The underlying assumption is that frequent consumption feedback makes consequences of energy behavior and energy behavior changes more immediate and salient.
The present situation in many countries is that consumers are infrequently informed by utility companies about their household energy consumption. In the Netherlands, for example, existing energy billing procedures are designed in such a way that consumers are charged with monthly or bimonthly advance payments based on an estimate of their current energy use. Once a year their energy meters are read by the utility company to determine their real energy consumption and a new estimate is computed for their advance payments. From a psychological point of view such a billing procedure is rather unfortunate, given the time lag between energy acts and the consequences of these acts.
Ellis and Gaskell (1978) have developed a valuable conceptional framework, drawing on previous work by Annett (1969), of cognitive factors underlying energy feedback. Using an information-processing model of consumer behavior (cf. van Raaij, 1977), these authors make a distinction between learning and motivational functions of energy feedback. The learning function refers to the process through which consumers learn by receiving feedback information on their task performance. In this sense, feedback teaches the consumer how the energy system in which the consumer is engaged operates. Through regular feedback the consumer may improve and increase his knowledge and understanding of this system, such as the relationship between energy behavior (e.g. thermostat setting, appliances use) and energy consumption. The motivational function refers to the likely possibility that feedback increases general awareness of energy issues and may raise the saliency of conservation goals. For instance, frequent feedback on financial savings from reduced energy consumption may result in explicit energy conservation goal setting by consumers.
Besides these two functions, feedback may also give consumers the sense of having greater personal control of their energy use (Stern & Gardner, 1980). In later studies Ellis and Gaskell reformulated their ideas about feedback, in the sense that the motivating and learning functions of feedback were no longer viewed as distinct processes but as different aspects of the same process (Gaskell & Ellis, 1982, p. 120).
A number of factors have been identified for energy feedback to be effective:

e.g. feedback must be given regularly, feedback must be specific enough to enable consumers to relate changes in energy consumption to changes in energy behavior, and the feedback source has to be perceived as a credible one (cf. Geller, Winett & Everett, 1982, p. 184).

Feedback may be seen as a potential psychologically effective intervention for promoting residential energy conservation. However, this does not necessarily imply that feedback is a feasible option from a policy point of view. It is doubtful, for example, whether frequent feedback (say on a daily basis) is a cost-effective policy measure. We will return to this problem in subsequent sections of this Chapter as well as in Chapter 4.

3.2.4 Self-monitoring of energy consumption

Self-monitoring is a special form of feedback that entails systematic self-observation followed by self-recording of those observations. Self-monitoring has been widely and effectively used in behavior therapy and has considerable potential as a self-control, behavior change procedure (cf. Kazdin, 1974; Nelson, 1977; Richards, 1977; Thoresen & Mahoney, 1974).
In essence, self-monitoring is a two-stage process: it enables subjects to determine that the target behavior has indeed occurred and subjects are to use the self-recording devices and procedures to monitor the occurrence of the target behavior. As such, self-monitoring has both an assessment and a therapeutic function (Nelson, 1977).
Self-monitoring has been applied in behavior therapy to influence a wide variety of undesired behaviors, e.g. cigarette smoking, alcohol drinking, drug abuse, overweight, poor study results.
Several authors have argued that self-monitoring of energy use could very well be a potentially effective behavioral intervention for promoting residential energy conservation (de Boer, 1982; Ellis & Gaskell, 1978; Geller, Winett & Everett, 1982; Winett, Neale & Grier, 1979). According to de Boer (1982) regular self-monitoring of residential energy consumption - through frequent meter readings - could have at least four important functions:

1. attention: regular meter readings direct continuous attention to household energy consumption which may stimulate energy conservation goal setting.

2. orientation: frequent self-recording of energy consumption may increase consumers' knowledge of factors determining household energy use. Consumers may consequently try to explain changes in energy use patterns as well as

make an effort to infuence these patterns through behavior changes.

3. discussion: regular energy meter readings may raise household discussions about energy consumption and may stimulate concern with energy conservation.

4. confirmation: frequent self-recording of household energy consumption may to some extent provide consumers with feedback on their conservation performance. This information may contribute to the institutionalization of new behavioral patterns.

Self-monitoring of residential energy consumption is based on the same psychological principles as feedback, in fact self-monitoring can be characterized as an active form of self-feedback. In this respect self-monitoring teaches the consumer how residential energy systems operate (learning function) and may influence the saliency of conservation goals (motivational function). Also, self-monitoring may increase consumers' personal control of their energy use.
From a policy point of view self-monitoring is an interesting intervention given its simpleness, its low costs and its large-scale applicability. At present, most utility companies in the Netherlands provide households in their service areas with self-monitoring recording charts, and the total number of circulating recording charts is estimated at some two million (SVEN Apeldoorn, oral information, December 1982). However, no empirical studies in this country are available on the effectiveness of self-monitoring in promoting residential energy conservation.

3.2.5 Energy conservation incentives

Two other strategies suggested by Platt to avoid or eliminate social traps include to add counterreinforcers and to add short-term rewards for competing behaviors. These strategies are in line with Olson's collective action theory from which the statement can be derived that the provision of selective stimuli is necessary to reinforce energy conservation by consumers.
Four possible procedures of behavior change could be mentioned in this context: positive reinforcement, negative reinforcement, positive punishment, and negative punishment. Definitions of these four procedures will be borrowed from Mikulas (1972, p. 87-88).
Positive reinforcement refers to a contingency between the onset of a pleasant event and a behavior that results in an increased frequency of the behavior. Examples are reduced energy bills as a consequence of conservation efforts, tax rebates for home insulation, monetary incentives for installing high-efficiency

boilers. Negative reinforcement refers to a contingency between the offset of an aversive event and a behavior that results in an increased frequency of the behavior. An example is reducing energy consumption following a high energy bill in order to avoid another one.

Positive punishment is a contingency between the onset of an aversive event and a behavior that results in a decreased frequency of the behavior. For instance, children who make punishing remarks to their parents in case of excessive energy consumption. Negative punishment is a contingency between the offset of a pleasant event and a behavior that results in a decreased frequency of the behavior. An example is replacing a decreasing block rate by an increasing block rate.

Behavior analysts prefer positive reinforcement for behavior modification over negative reinforcement and punishment for a number of reasons, including the fact that positive reinforcement elicits less reactance responses, is easier to administer, more cost-effective in the long run, and is more acceptable from a moral point of view (Bandura, 1969).

One has to realize that positive reinforcement can have manifold expressions: desired behaviors may be reinforced through material or immaterial rewards, and through individual or social rewards.

Both Cone and Hayes (1980) and Geller, Winett and Everett (1982) make the important observation that in spite of this preference for positive reinforcement, environmental policy shows a clear tendency for applying primarily negative reinforcement and punishment in promoting environmentally-conscious behavior, mainly through laws, fines and ordinances. As far as energy conservation policy is concerned there is a wide scope of possibilities to apply positive reinforcement principles, especially through offering monetary incentives contingent upon conservation behavior. Examples are: subsidies for home insulation, tax credits for the purchase of solar hot water or heating systems, monetary incentives for installing high-efficiency boilers, energy rate structures reinforcing conservation behavior, rebate strategies in master-metered dwellings, reduced energy rates during nonpeak periods.

In spite of the fact that the use of monetary incentives as a form of positive reinforcement of promoting residential energy conservation theoretically can well be founded on the basis of behavior modification theory and microeconomic theory, one has to recognize some possible negative side-effects. Some authors have argued that incentives may only be temporarily effective and may strengthen the individualistic attitudes at the root of the energy commons dilemma which inhibits a change toward proconservation attitudes (Stern & Kirkpatrick, 1977). Also, several authors have strongly doubted the cost-effectiveness of incentive schemes (Cook & Berrenberg, 1981; Stern & Gardner, 1978), and have raised equity issues (Ellis & Gaskell, 1978). An interesting critique has been put forward by Hayes (1976) who argues that financial rewards contingent upon energy conservation behavior may be invested in energy wasteful behavior, e.g. the purchase of a larger or second car. Hayes labels the possible occurrence of these unintended side-effects the "boomerang law of energy conservation" (Hayes, 1976, p. 63).

Summary

In this section a number of antecedent and consequence strategies (information, prompts, modeling, feedback, self-monitoring, incentives) were discussed which are or could be used to promote residential energy conservation. The dynamic character of and interaction between these two broad classes of interventions was underlined. Behavioral assumptions underlying these interventions were placed in the context of some theoretical considerations outlined in the preceding chapter.

It should be stressed that to a certain degree the interventions mentioned are only effective when applied in combination; energy consumption feedback, for example, will only be effective if consumers are aware of appropriate ways of energy conservation, which therefore illustrates the importance of conservation information.

3.3 Behavioral experiments in energy conservation: a review and critique

In this section the relatively new and growing body of behavioral studies on promoting residential energy conservation will be reviewed in some detail. (1) The main goals of this review are (a) to point out the strengths and weaknesses of these studies with respect to research design, intervention techniques, methodology, results, and conclusions, and (b) to derive guidelines - both in an empirical and theoretical sense - for this study.

A basic dimension of this review is to determine the energy policy relevance of the interventions evaluated in these experiments.

Two restrictions have to be made. First, since a number of reviews of behavioral approaches to residential energy conservation are already available (see e.g. Carlyle & Geller, 1979; Cook & Berrenberg, 1981; Ellis & Gaskell, 1978; Ester, 1979a; Geller, Winett & Everett, 1982; Joerges & Olsen, 1979; Lloyd, 1980; Stern & Gardner, 1980; Winett, 1980) this review will include conclusions from these previous studies. (2)

Second, behavioral experiments on energy conservation in master-metered settings (see e.g. Cook, 1978; McClelland & Cook, 1980a, 1980b; McClelland & Belsten, 1979; Newsom & Makranczy, 1978; Slavin & Wodarsky, 1977; Slavin, Wodarksy & Blackburn, 1981; Walker, 1979) are excluded from this review, given the specific energy characteristics of these settings. (3)

Our survey of behavioral experimentation on energy conservation will be structured according to the distinction between antecedent and consequence strategies that was discussed in the previous section.

3.3.1 Antecedent strategies and residential energy conservation

This section summarizes the results of experiments on the effectiveness of two antecedent strategies on encouraging residential energy conservation, namely information and modeling.

3.3.1.1 Information

One of the first behavioral experiments on the effectiveness of energy conservation information was a study by Heberlein (1975). The hypothesis guiding his experiment was that conservation campaigns have little or no impact on household energy consumption due to structural constraints and because of the generally poor relationship between attitudes and behavior. The study was conducted in the spring of 1973, just prior to the energy crisis, among 84 residents of six apartment complexes in Madison, Wisconsin. The target behavior was electricity consumption. Baseline consumption was assessed by surreptitiously reading subjects' electric meters -located in the buildings' basement - daily for 12 successive days.
Two experimental groups received written prompts to decrease their electricity consumption, one experimental group was prompted to increase (sic) their consumption. A control group received no prompts. The information in the experimental groups manipulated variables like beliefs about the costs of electricity, beliefs about the consequences to others of the use of electricity, and the personal responsibility of the consumer for these consequences.
The first experimental group received a letter outlining both personal and social negative consequences of excessive electricity consumption. The letter emphasized the consumer's personal responsibility for these consequences. In the second experimental group subjects received a pamphlet with energy-conserving tips. Subjects in the third experimental group were sent a letter emphasizing the beneficial aspects of electricity consumption (e.g. its low cost) and argued against blaming individual consumers for adverse consequences of electricity consumption. All subjects in the treatment groups were contacted by

telephone and urged to read the information.
Upon receipt of the information subjects's electric meters were read for 12 days.
A before-after analysis showed no significant differences between baseline and experimental electricity consumption. None of the experimental conditions turned out to have any effect on electricity use.
The 1973-74 energy crisis provided the possibility for conducting a natural experiment within the context of this study to evaluate conservation appeals, since numerous energy conservation campaigns were initiated. Exactly one year later subjects' meters were again read over a 24-day period. Again, no statistically significant differences in consumption data were obtained, indicating that the energy crisis and the large-scale conservation campaigns appeared to have no direct effect on conservation behavior.
In a study by Winett and Nietzel (1975) monetary incentives for reduced energy consumption were compared with energy conservation information. Subjects were members of 31 households in Lexington, Kentucky who volunteered for the study. Pre-experimental interviews showed that subjects were concerned about environmental and energy issues and already practiced a number of energy conserving behaviors. In order to obtain a baseline consumption measure, subjects' meters were read weekly for a 2-month period. The experiment took place in 1974. Subjects were matched according to their baseline consumption of electricity and natural gas and randomly assigned to an information or incentive condition. In the information condition subjects received a detailed manual on energy conservation, together with a self-monitoring energy-use recording sheet. This recording sheet enabled subjects to monitor consumption changes. They were instructed to read their utility meter weekly. In the incentive condition a payment schedule was in effect offering cash payments for reduced energy consumption (See Section 3.3.2.3). Subjects in this condition also received the conservation manual plus recording sheet. The experimental period lasted for four weeks. A no-treatment control group was not employed in this study.
The results show that the incentive group averaged approximately 15% more electricity reduction than the information plus self-monitoring group. No differences were reported for natural gas consumption. The between-group differences in electricity use tended to be maintained in a 2-week and 8-week follow up.
These results have to be interpreted carefully since no control for weather changes was employed in this study by using either a no-treatment control group or by applying the degree day method.
In a study by Palmer, Lloyd and Lloyd (1978) the effects of two types of prompts and feedback on electricity consumption were examined. Subjects were four (sic) households from Des Moines, Iowa who agreed to participate in the

study. Two types of prompts were used: 1) a series of eight typewritten prompts (e.g. "Kill-a-Watt, Conserve Electricity!") was taped to the inside of subjects' storm door, and 2) a personal letter from the Director of the Iowa Office of Energy describing the instability of electricity supplies and asking for a 20% reduction of electricity consumption. The two types of feedback consisted of either daily feedback on consumers' electricity use or daily feedback plus a report of the expected monthly bill was projected. The experimental conditions to which subjects were exposed included: daily feedback, daily feedback plus cost information, daily typewritten prompts, daily typewritten prompts plus feedback, government prompt.

A multiple-baseline-across-"groups" design was used in this experiment (see Palmer, Lloyd & Lloyd, 1978, p. 667-669). Data were collected in 1974 for a total of 106 days.

In general the results show some evidence for the effectiveness of prompting techniques. No clear differences were obtained between experimental conditions. The number of subjects and the short period participants were exposed to each of the treatment strategies seriously limit the external validity of this study.

Hayes and Cone (1977) studied the effects of conservation information, monetary incentives, and feedback on electricity consumption. Their study was conducted in the first half of 1975 among four (!) households from an 80-unit housing complex for married students at West Virginia University, Morgantown. Target behavior was electricity use for air conditioning mainly.

The design of the study combined multiple-baseline and withdrawal procedures in addition to a control group to establish the separate effects of the following conditions: 1) covert baseline, 2) overt baseline, 3) monetary incentives (a payment schedule was employed for reduced electricity consumption), 4) daily feedback, and 5) information (subjects received a poster with various tips on how to reduce electricity consumption).

The results of this experiment indicated that monetary incentives produced immediate and stable reductions. Feedback appeared to be the next most effective intervention, whereas information generally was ineffective or yielded only temporary effects. With regard to this last finding the authors remark that "curiously, information in the form of massive educational campaigns seems to be the main strategy adopted by governmental agencies and power companies to control the consumption of energy. Perhaps the money spent on such campaigns could be better spent in developing and implementing rebate or feedback systems" (Hayes & Cone, 1977, p. 433-434). However, as was the case with Palmer _et al._ (1978) study one has to realize that inferring such a conclusion from a study with only four subjects is questionable.

In a study by Winett, Kagel, Battalio, and Winkler (1978) the effectiveness of monetary incentives, feedback, and information was examined. Subjects were 129 volunteer households from College Station, Texas. Pre-experimental interviews suggested that subjects were generally well informed about conservation practices. The experiment was conducted in the summer of 1975, target behavior was electricity use. After a 2-week baseline period, subjects were randomly assigned to one of five groups: 1) high-rebate group (monetary incentives for reduced electricity consumption, plus weekly feedback, plus conservation information), 2) low-rebate group (same treatment as the high-rebate group but lower payments), 3) feedback group (weekly feedback plus conservation information), 4) information group (information about household energy conservation and about how to compute one's own electricity bill), and 5) control group. These treatments were in effect for four weeks.

It appeared from the results that both rebate groups produced the largest mean changes from baseline consumption (3.5% and 4.5%, respectively). Remarkably, both the information and feedback group showed an increase (7.25% and 1.75%, respectively) in electricity use. After four weeks experimental conditions were modified. Payments and feedback were withdrawn from both rebate groups, the information group received a high-rebate plan, and the control group received the former information condition booklets. These new conditions were also in effect for four weeks.

Results indicated that the former information-only group significantly reduced its electricity use (by 7.6%), supplying information to the former control group resulted in a nonsignificant increase. After withdrawing payments and feedback from both rebate groups, electricity consumption showed a continuing decreasing tendency (reductions of 3% and 5%, respectively).

Craig and McCann (1978a) applied a consumer information-processing approach to analyzing the problem of communicating energy conservation information to consumers. (4) The study was conducted in New York in 1976, the target behavior was electricity use. The main research question was whether source credibility and repetition of message are related to effectiveness of energy conservation prompts. Subjects were 1000 consumers from the service territory of Consolidated Edison of New York. A number of selection criteria was used to ensure that customers were likely to have air conditioning. Subjects were randomly assigned to one of four experimental groups and one control group.

The following treatments were used: 1) low-credibility source/one conservation message, 2) low-credibility source/repeated conservation message, 3) high credibility source/one conservation message, and 4) high-credibility source/repeated conservation message.

Restricting the presentation of the results to the effects of these experimental conditions on actual consumption of electricity, it was found that repetition of message had no effect but manipulating source credibility did: conservation messages from a high-credibility source resulted in significantly more electricity conservation than messages from a low-credibility source.

We will conclude this review of primarily U.S. studies on the effectiveness of energy conservation information by summarizing the results from two European studies.

Gaskell, Ellis and Pike (1980) used a self-monitoring and information strategy to influence the natural gas and electricity consumption of a sample of 160 London households. The experiment was conducted in the winter of 1979-80. Subjects were assigned to either one of three experimental groups or a no-treatment control group. The experimental conditions included: 1) self-monitoring by subjects on their energy consumption, 2) supplying subjects with an energy conservation leaflet, and 3) self-monitoring and conservation information combined. The experimental conditions were in effect for 8 weeks. Baseline measures were taken for 2 weeks.

It was found that self-monitoring alone did not lead to notable decreases in energy consumption, the information condition, however, produced significant savings in electricity (8%) and natural gas (9%) consumption. Self-monitoring and information combined yielded the largest reductions (11% and 22%, respectively).

In a Dutch study by Midden, Weenig, Houwen, Meter, Westerterp, and Zieverink (1982) four experimental treatments were studied with regard to both natural gas and electricity consumption: 1) weekly feedback plus conservation booklets, 2) comparative feedback (i.e. feedback on subjects' own energy consumption compared with consumption of other subjects in similar circumstances) plus conservation booklets, 3) monetary incentives plus comparative feedback plus conservation booklets, 4) conservation booklets only, and 5) control group.

Subjects were 82 households living in multi-family apartment dwellings in Voorschoten (near The Hague). The study was conducted in 1980, the experimental conditions were in effect for 12 weeks.(5)

Subjects were interviewed before and after the experiment. Baseline consumption was assessed by reading subjects meters during a three-week period prior to the experiment.

Linear trend analysis showed that as far as electricity consumption is concerned, individual feedback, comparative feedback, and monetary incentives were about equally effective in reducing consumption (approximately by 18%), conservation information alone did not significantly decrease consumption. For natural gas consumption this picture is somewhat different as here comparative feedback was considerably less effective (reduction of 6%) than individual feedback and monetary incentives (reductions of about 18%). Conservation information alone, again, was ineffective.

The general conclusion from these studies seems to be that communicating energy conservation information alone is not a very powerful instrument in promoting consumer energy conservation. In spite of the fact that providing conservation information is probably the most widely used governmental policy with respect to influencing consumer energy behavior, our review suggests that from a point of view of effectiveness such a policy is severely limited (cf. Carlyle & Geller, 1979; Ester, 1979a; Geller, Winett & Everett, 1982; Lloyd, 1980). The current conclusion is that "information alone does not work" (Winett, 1980, p. 338). Similar behavioral studies on promoting other environmentally conscious behaviors (e.g. recycling, litter control) tend to confirm this conclusion (Cone & Hayes, 1980; Ester & Winett, 1982; Geller, Winett & Everett, 1982).

It should be added, however, that these studies at the same time provided evidence for the notion that _prompting_ proenvironmental behavior does produce behavior change under certain conditions. The conditions appear to be that effectiveness of prompting is related among other things to its specificity and non-demanding character, as well as to the psychological convenience of the requested behavior (Geller, 1980).

3.3.1.2 _Modeling_

So far, only two studies (Winett, Hatcher, Fort, Leckliter, Love, Riley & Fishback, 1982; Winett, Leckliter, Love, Chinn & Stahl, 1983) are available which quasi-experimentally investigated the effectiveness of modeling in encouraging consumer energy conservation. The effectiveness of modeling in both studies was assessed by showing videotape modeling programs to consumers that demonstrated residential energy conservation strategies.

The first Winett _et al._ study consisted of a winter and a summer study. The winter (1980) study was conducted in Blacksburg, Virginia. Subjects were residents from two all-electric townhouse complexes randomly assigned to experimental groups across complexes. Two videotapes were produced. (6) Tape A (discussion tape) consisted of a staged interview program in which the seriousness of the energy crises was discussed by a male and female couple. No specific conservation practices were indicated. Tape B (modeling tape) showed - using again a couple as model - how people can successfully adjust to changing temperatures. Appropriate and inappropriate adjustment and conservation practices were indicated. The tape emphasized the positive consequences (material and immaterial) of conservation. Both tapes had an equal length: 20 minutes.

Subjects were 83 volunteers assigned to one of the five following treatments: 1) a feedback and discussion tape group, 2) a group receiving conservation information and the modeling tape, 3) a feedback and modeling tape group, 4) an information and discussion tape group, and 5) a control group.

The feedback information (daily) was almost similar to the information provided in the Winett, Neale, and Grier (1979) study. (7) After a three-week baseline period, subjects were invited to attend meetings for viewing the videotape. Next, subjects' electricity consumption was monitored for a period of 35 days. Results show that during the five-week intervention phase the groups receiving feedback and/or modeling tape showed about 12% less electricity use than the group receiving information and the discussion tape and about 17% less than the control group.

The summer (1980) study replicated a number of findings from the winter study but had a somewhat different design. The study was conducted in Salem, Virginia among 54 residents from two all-electric apartment complexes. Three experimental treatments were in effect and one control group. Experimental treatments included: 1) a feedback and modeling tape group, 2) a conservation information and feedback group, and 3) a conservation information and modeling group.

The modeling tape in this summer study showed subjects a number of alternative practices for remaining comfortable (e.g. by natural ventilation, proper use of fans) demonstrated by a male and female couple, while at the same time reducing their electricity consumption for air conditioning use.

Results indicate that during the intervention period (30 days) the feedback and modeling tape group showed a reduction of 22% in their electricity consumption, the information and feedback group reduced their consumption by 19%, and the information and modeling group by 12% compared to their baseline consumption.

The second Winett _et al._ study both replicated and extended the first modeling experiment. Again, a winter and summer experimental field study was done. The studies attempted to assess the effects of videotape modeling with and without support and interaction, in a group and home environment, and as delivered over a cable TV system. Both studies were conducted in Roanoke, Virginia.

The winter (1982) study was conducted among 80 residents from two all-electric apartment complexes. Experimental conditions consisted of 1) group - support, 2) group - no support, 3) home - support, and 4) home - no support. Also a non-treatment comparison group was used. In the group - no support condition at least one member from assigned households attended a group meeting which primarily entailed viewing a 26-minute videotape about household energy conservation strategies. The group - support condition was similar to the group - no support condition except that showing the video program was followed by small group discussions about the demonstrated conservation strategies. The home - no support condition was the same as its group counterpart except that the videotape was played in subjects' home on there TV using a video cassette player. The home - support condition, finally, followed the home - no support condition except that the demonstrated conservation strategies were discussed with subjects after viewing the video program. Conditions were repeated after about two weeks but with an abbreviated 16-minute videotape program that reviewed and summarized the conservation strategies showed in the first program. Again, hygrothermographs were placed in about two-thirds of the households to continuously measure temperature and humidity. After a 3-week baseline period, the interventions were in effect for 6 weeks. Findings indicated, that the videotape program was effective irrespective of experimental condition. Subjects reduced their overall household electricity consumption by between 9% - 14% and 25% on heating. Consistent with results from the first study it was again found that subjects could live comfortably at lower temperatures.

The summer (1982) study was conducted among 150 residents from a large subdivision of single detached homes which were subscribers to the local cable TV system. Five groups were involved in the study to assess the effects of a videotape program on household energy conservation strategies: 1) no-contact - control group, 2) contact - control group, 3) no-contact - media group, 4) contact - media group, 5) contact - media - home-visit group.

The no-contact - control group only had its outdoor electricity meters read, similar to the contact - control group which, however, also completed weekly clothing and comfort forms and were partly equipped with hygrothermographs. The no-contact - media group was the same as the no-contact - control group except that subjects were prompted to watch the TV program. The contact - media group was similar to the no-contact - media group except for the addition of the same form completion procedure as the contact -control group. The contact - media - home-visit group finally, was the same as the contact - media group except that after the program viewing, subjects were visited by a staffperson to discuss, explain and work out conservation strategies for their household. The 20-minute video program was showed four times. (8) The study contained a 3-week baseline period and a 5-week summer intervention period, as well as some follow-up phases.

Results showed that the no-contact - media group reduced its electricity use by 11% to its respective control group, while the contact - media group reduced by 7% and the contact - media - home-visit group by 8%. Findings suggest that the TV program alone and not personal contact was the effective element. Again, no loss of comfort was reported.

Given the fact that those are the only two studies on modeling with respect to consumer energy conservation it is, unfortunately, not possible to compare these findings with outcomes of other studies.

3.3.2 Consequence strategies and residential energy conservation

In this section a number of behavioral experiments will be reviewed which tested the effectiveness of consequence strategies in promoting residential energy conservation. First, experiments will be analyzed which tested self-monitoring (Section 3.3.2.1), and next results of studies will be summarized which applied feedback strategies (Section 3.3.2.2). Finally, experiments which used monetary incentives to reduce consumer energy consumption (Section 3.3.2.3) will be discussed.
Outcomes of studies of which the research design has already been outlined in the previous sections will only be dealt with briefly.

3.3.2.1 Feedback

Feedback has been the most widely investigated behavioral intervention for changing consumer energy consumption. (9) The previously discussed studies by Hayes and Cone (1977), Midden, Weenig, Houwen, Meter, Westerterp, and Zieverink (1982), Palmer, Lloyd, and Lloyd (1978), Winettt, Hatcher, Fort, Leckliter, Love, Riley, and Fishback (1981), Winett, Kagel, Battalio, and Winkler (1978), and Winett, Neale, and Grier (1979) all provided evidence that feedback is an effective intervention for promoting energy conservation.
Seaver and Patterson (1976) selected 180 households from a local fuel-oil distributor's list of continuing accounts in Pennsylvania. The study used a randomized treatments-by-blocks design with the following treatments: 1) no-feedback control, 2) informational feedback, and 3) informational feedback plus social commendation. Households were ranked by their oil consumption rate and were divided into 20 blocks of nine households with each block having similar consumption rates. Within blocks households were randomly assigned to one of these treatment conditions. The study was conducted from February to May, 1974. Two consecutive oil deliveries (in gallons) to each household were monitored. The first delivery was used as the informational feedback, the second delivery served to compute oil consumption following the experimental treatment. In two experimental conditions households received, together with the normal delivery ticket, information about their current consumption of oil, their oil consumption during a similar period in the previous winter, the percentage decrease or increase, and the resulting monetary savings or losses. In the feedback plus social commendation condition households which had reduced their oil consumption compared to the last winter also received a decal with the words "We Are Saving Oil". The control group only received the normal delivery ticket. The number of gallons of oil consumed during the second delivery period served as

the dependent variable.
Results indicate that all groups reduced their oil consumption, the control group by 9%, the feedback-only group by 17%, and the feedback plus commendation group by 22%. It has to be added that the difference between the control group and the feedback-only group was not statistically significant, in contrast with the difference between the control group and the feedback plus commendation group.
One might agree with Carlyle and Geller (1979) that one problem with this study is that an educational strategy was missing: households were not informed about possible conservation strategies.
Seligman and Darley (1977) also investigated the effectiveness of feedback in reducing residential electricity consumption. Their study was conducted in 1975 in a community in central New Jersey among residents of 29 physically identical townhouses. (10) Air conditioning use was the main target behavior of the study. Households were randomly assigned to either a feedback or a control group. Baseline consumption was assessed during a five-week period prior to the experiment. A regression line was plotted to predict daily electricity consumption from daily average temperature. Households in both conditions were asked to cut back on their electricity consumption by reducing their air conditioning use. The feedback group received daily feedback expressed in a ratio of actual over predicted electricity consumption. This treatment was in effect for four weeks.
Results showed that the feedback group used 10.5% less electricity than the control group, a difference which was statistically significant.
Becker (1977) combined goal setting and feedback as treatments to reduce household electricity consumption. The study was conducted in the summer of 1976 in the same community in New Jersey as the Seligman and Darley (1977) study among residents of 100 physically identical townhouses. Households volunteered for the study and were randomly assigned to one of five groups: 1) 2%-reduction goal plus feedback, 2) 2%-reduction goal, no feedback, 3) 20%-reduction goal plus feedback, 4) 20%-reduction goal, no feedback, and 5) no treatment control group. In the goal setting conditions subjects were asked to reduce their electricity consumption by either 2% (easy goal) or by 20% (difficult goal). Feedback was based on the ratio of predicted minus actual daily electricity consumption over predicted consumption and related to the reduction goal. Air conditioning use was again the target behavior. The experimental conditions were in effect for about three weeks. Information sheets were handed out indicating electricity consumption of various appliances.
It was found that the only group that used significantly less electricity than the control group was the 20%-reduction goal plus feedback group: a reduction

of 13% was achieved. Thus, task performance was best in a situation of a difficult goal combined with regular feedback on task performance. (11)

In another study on air conditioning usage by Becker and Seligman (1978) prompting and feedback techniques were used to reduce residential electricity consumption by air conditioning. The study was conducted in the summer of 1977 in the already mentioned community (Princeton) in New Jersey among 40 households living in similar townhouses. Households volunteered to participate in the study. Residences were randomly assigned to one of four conditions: 1) signalling device only, 2) signalling device plus feedback, 3) feedback only, and 4) control group. The signalling device consisted of a 3.8-watt light bulb - attached to the wall - which would blink repeatedly when the air conditioning was on and when the outside temperature was below 68°F. The only way to stop the blinking was to turn off the air conditioning. Thus, a simple physical technology was used to indicate a situation of clear waste of electricity. Feedback was given three times a week and was calculated by subtracting actual consumption by predicted consumption for that period. Experimental conditions were in effect for about one month. Results demonstrated a significant effect due to the signalling device: households with the device used about 16% less electricity than households without the device. No significant feedback effect or interaction effect was observed. The authors estimate a payback period for the device of about two years.

Postexperimental interviews revealed a possible explanation for the ineffectiveness of feedback: subjects ignored the information because they reported little relationship between the feedback information and their conservation efforts. (12) This finding supports the notion that credibility of feedback is related to its effectiveness in generating conservation actions (cf. Craig & McCann, 1978a).

McClelland and Cook (1979) also tested a signalling device providing consumers with feedback on their electricity consumption. Their study was conducted in a development in Carrboro, North Carolina among 101 all-electric single-family homes with identical energy conservation construction packages. Twenty-five homes were equipped with a "Fitch energy monitor", a feedback device which provides residents with continuous information on their cents per hour electricity consumption. This information is displayed by light-emitting diodes on a panel inside the home. The independent variable in this study is the presence of the feedback monitor. Electricity consumption records were obtained for 11 months, from September 1976 to July 1977. Confounding variables (e.g. family size, home size) were removed statistically with multiple regression analyses. Results indicated that residents with feedback monitors used less electricity than residents without monitors. The differences averaged about 12%. Based on this reduction percentage the energy monitors would have a payback period of about 1.9 years.

One of the shortcomings of the studies reviewed is that no data are reported on the relationship between, for instance, responsiveness to feedback procedures and income levels. One of the exceptions is a study by Winett, Neale, Williams, Yokley, and Kauder (1978). This study was done in Greenbelt, Maryland in 1977 among residents from three different types of residential areas. These residential areas corresponded with different income levels and residential structures (e.g. physical structure of homes, household size). Subjects were 76 volunteer households from these three areas (N = 36, N = 11, N = 29, respectively). Control groups were used in each area (N = 22, N = 10, N = 14, respectively). A cluster procedure was used to assign households within each area for either 1) an individual feedback condition, 2) a group feedback condition, and 3) an individual plus group feedback condition. Each household set its own conservation goal (reduction percentage). After baseline data collection group meetings were held to explain the feedback procedures. Subjects received daily written feedback about the total kilowatt-hours consumed the preceding day (expressed in terms of individual households, groups of households, or both) and a percentage comparison with expected use. The conditions were in effect for five to six weeks.

Results show that the greatest reductions (about 20%) were found in the individual plus group feedback condition. Group feedback-only yielded negligible reductions. The study provided some support for the notion that responsiveness to feedback is positively related to income. (13)

Katzev, Cooper and Fisher (1981) studied the effects of feedback and social reinforcement on household electricity use. Subjects were 44 tenants of an apartment complex in Milwaukee, Oregon. The study was conducted in 1977. After a 2-week baseline period subjects were matched in groups of four according to similarity of electricity use. The four apartments in each of the eleven matched quads were randomly assigned to the following experimental conditions: 1) daily contingent feedback (KWH use during the previous day compared with subject's use the day before - translated into monetary costs - to the control group's average consumption for the preceding day), 2) three-day contingent feedback plus decal (same format as the daily feedback group but based on a three day period; if subjects showed lower consumption rates they also received a stick-on decal "We Are Conserving Electricity"), 3) three-day noncontingent feedback plus decal (regardless whether or not subjects reduced their electricity consumption they received a feedback slip indicating that they had been successful in conserving electricity during the preceding three-day period; they also received the decal), and 4) a no-treatment control group.

Experimental conditions were in effect for two weeks. Results indicate no significant changes or differences in electricity consumption neither between nor within groups. It is therefore concluded that in this study feedback had very little impact on electrical energy consumption.

Bittle, Valesano and Thaler (1979) investigated the effectiveness of daily cost feedback in reducing residential electricity consumption. The study was conducted in the summer of 1976 in a rural Southern Illinois community among 30 volunteer middle income families. The study utilized a within-group reversal design with a delay group. Following a 12-day baseline period, 15 families (Group A) received daily (written) feedback for a 42-day period. Next, a 24-day reversal period followed in which daily cost feedback was given to the other 15 families (Group B), no feedback was provided to Group A. Subjects were randomly assigned to either of these groups. Feedback provided written information about date, number of kilowatts consumed, cost of kilowatts used, and cumulative cost of electricity since beginning of treatment.

Results show that during the first feedback period Group A families used an average of 4% less electricity compared to the no-feedback control Group B. It appeared, however, that during the second feedback period, after a short reversal of consumption patterns, subjects <u>not</u> receiving feedback (Group A) used 6% to 21% less electricity than subjects receiving feedback (Group B). A further inspection of the data revealed that in periods with comparable temperatures electricity consumption was lower during the feedback period. It is found that feedback is most effective when the need for electricity for cooling purposes is lowest. The finding that reversal of experimental conditions did not result in reversal of consumption patterns is explained in terms of possible carry-over effects.

Once again, this study illustrates the importance of proper research designs (e.g control groups) and appropriate correction measures for outside temperature in behavioral energy experiments.

The final study that will be described in this section is an experiment by Hayes and Cone (1981). Their study is of particular interest because of their criticism on mainstream behavioral experimentation on consumer energy conservation. One of their critical notions is that feedback frequencies commonly tested are impractical from a policy point of view. This is especially true for daily feedback. Their suggestion is to investigate the effectiveness of lower feedback frequencies, preferable in connection with already exisiting "natural" feedback channels. One of these channels is the monthly utility bill (U.S.) consumers receive. The study was conducted in the first half of 1976 among 40 residents of Pawtucket, Rhode Island. Subjects were nonvolunteer participants matched to a sample of volunteers from an earlier study (Hayes, 1977) and ran-

domly assigned to either a baseline-only control group or a monthly feedback group. The study was designed in an ABA fashion. There were two baseline periods: one over the years 1973 and 1974, and a second period during the year prior to the start of the experiment (February, 1976). The experimental treatment lasted for four months. The feedback group received a letter each month shortly after their utility bill providing information about percent change in electricity consumption over the same month during baseline (1973, 1974), the number of kilowatthours, and actual dollar amounts involved.
Results indicate that the feedback group decreased its electricity consumption relative to baseline consumption by about 5%, whereas the control group showed an increase of 2%. Withdrawal of experimental treatment resulted in an 11% increase of electricity consumption by the feedback group, again compared to baseline consumption.
The authors conclude that these relatively clear effects of this feedback procedure could be easily implemented by simply adding the feedback information to the monthly utility bill. Thus, from a policy point of view this study seems to be rather promising, also given the fact that use was made of a nonvolunteer sample.

3.3.2.2 Self-monitoring

As to our knowledge only three studies (Gaskell, Ellis & Pike, 1980; Winett, Neale & Grier, 1979; Winett & Nietzel, 1975) exist which examined the effectiveness of regular self-monitoring by consumers of their own energy consumption in generating possible conservation efforts. This observation is quite remarkable in itself, given the fact that self-monitoring at face value appears to be a relatively inexpensive and large-scale applicable intervention.
As was summarized in Section 3.3.1.1, the Gaskell et al. (1980) experiment showed that self-monitoring alone was ineffective to produce conservation efforts but self-monitoring and conservation information combined resulted in reduced natural gas and electricity consumption. The self-monitoring treatment consisted of daily meter reading recording by subjects on specially prepared charts. The charts allowed for a graphical plotting of daily natural gas and electricity consumption and conversion into daily and weekly cost. Subjects could also record their daily use of appliances as well as mean daily outside temperature.
The Winett et al. (1979) study was conducted in a suburban Maryland upper-middle class all-electric townhouse community near Washington, D.C. in winter 1978. Subjects were 42 volunteer households randomly assigned to either a feedback, a self-monitoring, or a control group.

The feedback group received daily feedback on their electricity consumption providing information about the households' prior day's electricity consumption, the percentage decrease or increase from a baseline measure and from a reduction goal (conservation percentage) chosen by each household, and, finally, an estimate of the households' monthly electricity bill in dollars. The self-monitoring group was taught how to read their (outside) electricity meters and received four weekly meter reading recording forms. Subjects were instructed to record their KWH use every day. Participants also received daily notes indicating their expected use for the prior day as predicted from baseline consumption measures corrected for outside temperature.
Subjects in both experimental conditions were given written conservation information. After a three-week baseline period, the experimental conditions were in effect for 28 consecutive days.
Results indicate - using a control group comparison - that both experimental groups reduced their electricity consumption significantly. It turned out that feedback was more effective than self-monitoring (reductions 13% and 7%, respectively). Two follow-up periods showed that the obtained reductions were maintained.
As far as the self-monitoring condition is concerned the authors conclude that "with minimal training and prompting, consumers, who before the study had never read their electricity meters, could become highly persistent and reliable meter readers" (Winett, Neale & Grier, 1979, p. 182). This conclusion together with the relatively inexpensive character of self-monitoring induces the authors to make a plea for further experimentation with this strategy.
The final study which used self-monitoring as an experimental treatment to influence consumer energy use is the already described experiment by Winett & Nietzel (1975) which compared monetary incentives for reduced consumption with self-monitoring plus conservation information. It turned out that both conditions were effective but a between-group analysis indicated that self-monitoring was the least effective intervention. Again, it has to be emphasized that no control was used for weather fluctuations and that the outcome only held for electricity consumption and not for natural gas use.

3.3.2.3 Monetary incentives

A number of studies examined the effects of monetary incentives on household energy consumption. In the earlier (Section 3.3.1.1) described study by Winett and Nietzel (1975) cash payments were made contingent on reduced electricity and natural gas use by residential consumers. Subjects who reduced their weekly energy consumption by 5-10% below baseline use received a $2 payment; reductions of 11-20% resulted in a $3 rebate; and reductions of more than 20% earned $5. Bonus payments were given to households averaging the greatest reductions. As outlined before, this incentive condition was compared to an information condition. Results show that the incentive group reduced its electricity consumption by about 15% more than the information group. No significant differences were observed for natural gas consumption.

Again, it has to be emphasized that the absence of a no-treatment control group and of weather control measures (degree days) limit the reported findings.

In the Hayes and Cone study (1977) a similar payment schedule was used: a 10-19% reduction earned a $3 payment; 20-29% earned $6; 30-39% earned $9; 40-49% received $12; and 50% or more resulted in a $15 rebate. These payments were reduced in the course of the study but the basic schedule remained the same. As outlined earlier, a rather complex design (multiple baseline plus withdrawal procedures) was used.

Results demonstrate that payments produced stable reductions in electricity consumption, the 100% payment condition being most successful by yielding an average reduction of 34%.

The treatments in the Winett, Kagel, Battalio, and Winkler (1978) study included - besides an information and feedback condition - a high- and low-rebate group. The high-rebate group was eligible for weekly rebates of 30 cents for each 1% reduction in weekly electricity consumption below baseline use with a maximum of $15, plus an additional bonus of $10 if their percentage reduction in a 4-week period was among the largest half of households in the payment group. The low-rebate group received the same treatment, except that they were entitled to a 1.3 cent payment for each 1% reduction and a bonus of $2 only. The rebate differences amounted to price changes of 240% and 50%, respectively. Results demonstrate that only the high-rebate group significantly reduced its electricity consumption by about 12% over the course of the study.

As one remembers the Dutch study by Midden, Weenig, Houwen, Meter, Westerterp, and Zieverink (1982) also included a monetary reinforcement contingency. An explicit "equity principle" formed the basis for the assessment of the weekly payment: subjects with higher energy consumption rates had to reduce their energy use more than subjects with lower consumption rates in order to receive the same payments. Maximum weekly payment for natural gas reduction was Dfl. 40, maximum payment for electricity reduction Dfl. 7.50 per week. Results indicate that payments were effective in reducing energy consumption (by approximately 18% for both natural gas and electricity), but not necessarily more effective than feedback.

Finally, a study by Winett, Kaiser, and Haberkorn (1977), which has not been discussed before, also investigated the effects of monetary incentives on resi-

dential energy (electricity) consumption. The study was conducted in 1976 over a period of six weeks in an apartment complex in Kentucky. Subjects were 12 volunteer households randomly assigned to either an incentive or a control group. The incentive group received a similar payment schedule as the high-rebate group in the Winett, Kagel, Battalio, and Winkler (1978) study, together with daily feedback and energy conservation information.
After one week the payment schedule was reduced (by 50%) for half of the experimental households, the other half received zero payments. After another week experimental households received feedback only. Given the short duration of the different conditions the experimental results are difficult to interpret. The greatest consumption reductions (up to 31%) occurred during the high-rebate condition; feedback alone yielded a reduction of about 15%.
In summary, one may conclude that monetary incentives are effective in reducing residential energy consumption, though findings are not always consistent partly due to design difficulties. One has to realize, however, that the rebate schedules used often amount to price changes of several hundred percent! (Winett, 1980). Therefore, the policy applicability of such schedules may be disputed.
The general trend of the effects of monetary incentives seems to be in line with the outcomes of many studies investigating these incentives for promoting environmentally conscious behavior (e.g. litter control, recycling, mass-transit system use) (See Cone & Hayes, 1980; Ester, 1979a; Geller, Winett & Everett, 1982 for reviews).
In the next section some limitations of this approach will be discussed.

3.3.3 Conclusions and some critical notions

Considerable space has been taken to discuss and review behavioral experiments on residential energy conservation, especially with respect to research design, sample, intervention techniques, duration of experimental periods, methodologies, and results. The main reason for this extensive review was that by doing so both strengths and weaknesses of the studies concerned become clear, more clear at least than by simply summarizing major findings. In addition, the advantage of this procedure is that guidelines can be derived for designing our own experiment.
This section will summarize the existing state-of-the art of behavioral experimentation on residential energy conservation and will contain some critical notions with regards to the studies reviewed.

1. The main conclusion is that in general antecendent interventions (informa-

tion, prompts) are less effective - or better: mostly ineffective - than consequence interventions (feedback, self-monitoring, monetary incentives) in reducing residential energy conservation. It has to be noted, however, that antecedent interventions are often very poorly designed and restricted to leaflets and brochures, whereas much more creativity is put into designing consequence strategies. Thus, in a sense, self-fulfilling prophecies have been generated: antecedent interventions are not believed to be very powerful in changing energy behavior, therefore not much effort and research creativity is spent on its design, and consequently it is found that they are ineffective. This argument not only holds for behavioral energy research in particular, but for environmental research in general. Elsewhere we developed a more elaborate explanation of this argument (see Ester & Winett, 1982; Winett & Ester, 1983).

2. Although most studies on antecedent interventions have shown the ineffectiveness of energy conservation information through leaflets and brochures, little is known on why antecedents are ineffective. In general the studies reviewed did not investigate whether subjects were in need of conservation information, whether they used the information, how they evaluated the information, and whether the information matched their possible information needs. These shortcomings are important, since the dominant issue in communications theory is no longer what information does to people, but what people do with information (de Boer, 1982; Rogers & Kincaid, 1981).

3. Modeling, also an antecedent intervention but until recently hardly used in mainstream behavioral energy research, has proven to be effective in promoting residential energy conservation. Given the increasing importance of mass media like television, cable television, video etc., more experimentation on modeling is desirable to establish its usefulness for encouraging conservation.

4. Feedback has quite consistently proven to be an effective intervention in reducing consumer energy use. It has to be realized, however, that the feedback frequencies studied (especially daily feedback) have a disputable status with respect to policy applicability given cost-benefit considerations.

5. Self-monitoring, though hardly being investigated, has also shown to be capable of promoting conservation efforts. This finding is of interest in view of the obvious low-cost nature of this intervention.

6. Monetary incentives contingent on conservation behavior have been found to be capable of generating residential energy conservation. Again, it has to be

added that the rebate schemes were often quite unrealistic from a policy point of view.

The following conclusions and remarks pertain to the studies themselves.

7. Most studies were directed at promoting electricity conservation. Just a few studies are available on natural gas conservation. Given, for instance, the urgent need and importance of conservation of natural gas resources in the Netherlands, this has to be regretted.

8. Most studies use volunteers as subjects. An obvious problem is that volunteers may very well be a special consumer segment with respect to energy conservation (e.g. proconservation minded). (14)

9. A serious shortcoming of many studies reviewed is the small number of subjects (e.g. Hayes & Cone, 1977; Palmer, Lloyd & Lloyd, 1978), which severely limits the external validity of the findings due to the influence of chance factors.

10. Next, there seems to be a tendency to focus on middle-class households. One might hypothesize that these households do respond more positively to antecedent and consequence interventions as used in the studies reviewed than lower or upper-class households.

11. Another conclusion is that the period the experimental conditions were in effect in the experiments described is often of a rather doubtful short duration. A common range is between two and four weeks.

12. In addition to the preceding conclusion it is found that the same applies for the period in which baseline measures were taken. This implies that the influence of chance factors can hardly be controlled for.

13. For a number of studies the argument holds that a correct interpretation of findings is seriously hampered by either design imperfections, e.g. lack of no-treatment control groups (Winett & Nietzel, 1975), carry-over effects (Bitlle, Valesano & Thaler, 1979), or lack of clarity in baseline data treatment, e.g. correction for outside temperature, physical differences between dwellings (Gaskell, Ellis & Pike, 1980).

14. All U.S. studies used locations with dwellings provided with outside energy

meters. The unmistakable advantage of this procedure is obviously that energy consumption data can be measured unobtrusively, and thus eliminating possible attention effects. However, outside energy meters are a relatively rare phenomenon in the residential sector in the Netherlands. This fact creates specific research difficulties (e.g. attention effects). The difference in energy meter location presumably implies that findings from these studies - especially the ones on feedback - cannot be simply generalized to the Dutch situation.

15. It is often not clear from the studies reviewed whether conservation efforts caused by the experimental interventions persisted after the withdrawal of these interventions. Therefore, it is unknown to what degree the behavioral manipulations produced stable and long-term effects.

16. Next, there appears to be too little emphasis on conservation practices and energy behavior changes that are likely to endure (c.q. retrofitting).

16. Finally, a strong affinity and preoccupation of many U.S. researchers with rather traditional behavioristic paradigms often prevents them from paying attention to more cognitive variables. The neglect of such variables is reflected in the fact that quite often subjects' knowledge of residential energy matters, attitudes toward residential energy conservation and energy behaviors before and after implementation of the experimental interventions are not being investigated, at least not in a sophisticated way.
Because these cognitive variables probably function as mediating factors between the behavioral interventions and their effectiveness, and because these interventions may as such increase knowledge, or evoke attitude and/or behavior change, their neglect seriously limits our understanding of psychological factors affecting residential energy conservation.

Notes

(1) Portions of this section were adapted from Ester, P., Methoden ter bevordering van milieuvriendelijk en energiebewust consumptief gedrag, Vrije Universiteit, Insituut voor Milieuvraagstukken, Amsterdam, 1979a.

(2) See also: Anderson and McDougall (1980), Frankena (1977), Joerges (1979), McDougall and Anderson (1982).

(3) Studies of peaking behavior are also excluded from this review (see e.g. Heberlein & Warriner, 1982; Kohlenberg, Philips & Proctor, 1976; Blakely, 1978).

(4) See also Craig and McCann (1978b).

(5) See Midden, Meter, Weenig, and Zieverink (1981) for an English version of the results of this study.

(6) During my stay (1981) as a visiting Fulbright scholar at the Department of Psychology, Virginia Polytechnic Institute and State University, the senior author Richard Winett was kind enough to show me both videotapes.

(7) See for further details of the experimental set-up Winett, Hatcher, Fort, Leckliter, Love, Riley, and Fishback (1982).

(8) The videotape program was shown in a presentation by Richard Winett at the Institute for Environmental Studies, Free University, September 24, 1982.

(9) Some researchers even suggested that, if possible, feedback should be part of any behavioral program on residential energy conservation (Geller, Winett & Everett, 1982).

(10) This study is part of the Princeton University Twin Rivers project (see Socolow & Sonderegger, 1976).

(11) Becker refers with respect to this outcome to the work of Locke (1966, 1967, 1968) on the relationship between goal difficulty and task performance. See also Ilgen, Fisher and Taylor (1979), and Locke, Saari, Shaw and Latham (1981).

(12) See for differences in feedback calculation: Becker (1977), Becker and Seligman (1978), and Seligman and Darley (1977).

(13) For an excellent review of the relationship between responsiveness to feedback and rebates at the one hand, and income, energy expenditures, and energy budget share at the other hand, see Winkler and Winett (1982).

(14) Some studies, however, did not find major differences in energy use between volunteers and nonvolunteers (Winett, Neale, Williams, Yokley & Kauder, 1978; Hayes & Cone, 1981).

4. GUIDELINES FOR THIS STUDY: SELECTED BEHAVIORAL INTERVENTIONS, DESIGN REQUIREMENTS AND RESEARCH HYPOTHESES

4.1 Introduction

This chapter will first of all outline which guidelines directed the design of this experiment (Section 4.2). These guidelines will be derived from the extensive review of behavioral experimentation on consumer energy conservation. Next, the main research hypotheses guiding this study will be summarized (Section 4.3). As specified in Section 1.4 the research problem underlying the empirical part of this study was described as the empirical effectiveness of behavioral interventions in promoting residential energy conservation.

4.2 Selected behavioral interventions and design requirements

In the preceding chapter a number of important shortcomings of mainstream behavioral experimentation on consumer energy conservation were analyzed. These shortcomings seriously limit our understanding of behavioral interventions aimed at promoting residential energy conservation. In this section some guidelines will be described which directed the attempt made in this study to overcome these shortcomings. The next chapter will contain a more detailed description of the practical consequences of these guidelines for the experimental design.

4.2.1 Behavioral interventions to be tested experimentally: information, feedback, and self-monitoring

The two principal starting-points for the selection of behavioral interventions to be tested in this study are:

a. the interventions should have direct policy relevance

b. both antecendent and consequence interventions have to be included

In Chapter 3 a number of antecedent and consequence behavioral interventions was looked at in some detail: information, modeling, feedback, self-monitoring, and monetary incentives. It was argued that these interventions differ in many respects, for example with regard to policy relevance, and equity considerations. The selection procedure for interventions to be included in this study was based on the following viewpoints.

The inclusion of an energy conservation information intervention seems almost self-evident given the fact that it is widely applied by governments and utility companies, its relatively low cost nature in view of the potential large-scale application of informational strategies, and the observed phenomenon of consumer energy illiteracy. Although most studies reviewed concluded that informational strategies are generally ineffective, it can be argued that a more strict application of general principles from communications theory (e.g. target group segmentation, clear focus on salient beliefs, special attention to readability and comprehensibility of the information distributed) may considerably improve the design of these strategies and thereby yielding greater effectiveness (cf. Ester & Winett, 1982). Also, at least usually, energy conservation information will be indispensable when other behavioral energy conservation interventions are applied. Thus, an informational strategy will be included in this study.

Modeling, though a potential effective intervention, will not be tested mainly because of practical considerations and budget constraints.

Frequent feedback has quite consistently proven to be an effective intervention for promoting residential energy conservation. It has to be realized, however, that from the point of view of utility companies frequent (e.g. daily) feedback is simply not a feasible option given cost considerations. Another practical argument against frequent feedback, at least when provided through utility companies, is that in the Netherlands - unlike the U.S. - most residential dwellings have inside energy meters which hampers highly frequent meter recordings by utility personnel. Given the importance of motivational and learning functions of feedback (See Section 3.2.3), feedback will be tested in this study but the above-mentioned arguments resulted in a choice for less frequent schedules than normally studied: biweekly and monthly feedback. The earlier described study by Hayes and Cone (1981) using monthly feedback (See Section 3.3.2.1) provides additional support for this choice.

Self-monitoring by consumers of their household energy use will also be selected as a behavioral intervention to be tested in this study considering the fact that self-monitoring is largely based on the same psychological mechanisms as feedback, although the consumer is more actively involved. Also with respect to cost-benefit relations, self-monitoring is obviously a highly policy-friendly intervention.

In spite of the theoretical validity of many assumptions underlying the use of monetary incentives contingent upon decreased energy consumption, some considerations have led to the exclusion of this intervention in this study. As noted in Section 3.3.2.3 cost-benefit relations of monetary incentives are very disputable and often amounted in the studies reviewed to price changes of several hundred percent. Therefore, the policy relevance of monetary incentives is not immediately clear. "The studies of incentive effects of conservation have not demonstrated strong immediate effects in relation to program costs" (Stern & Gardner, 1980, p. 6). Another argument against using monetary incentives for decreased residential energy consumption, at least in the Netherlands, is that increasing block-rate energy pricing may create serious equity problems or - with individualized baseline consumption levels - administrative difficulties (Groenewegen, 1980). (1)

This selection procedure has resulted in a choice for testing the following behavioral interventions:
1. energy conservation information
2. biweekly energy consumption feedback
3. monthly energy consumption feedback
4. self-monitoring of energy consumption

The next chapter will contain a detailed description of how these interventions were designed, implemented and tested in this study.

4.2.2 Design requirements

Chapter 3 revealed a number of methodological shortcomings and problems in many behavioral experiments on consumer energy conservation. In order to overcome these shortcomings and problems, this section will briefly formulate some elementary design requirements for this experiment.

Unlike most U.S. studies, behavioral interventions in this experiment will be directed at conservation of both residential electricity and natural gas consumption. The reason behind this requirement is the fact that in the Netherlands space heating (mainly by natural gas) is responsible for more than 70% of residential energy use.
Given the intended policy applicability of the results, this experiment will not make use of a volunteer sample. In addition to this requirement, quite a larger number of subjects will be included compared to most U.S. experiments which both increases the external validity of this study as well as facilitates a sharper analysis of the hypotheses (See next Section) guiding this study. A basically similar argument has led us to select a less homogeneous sample than most experiments reviewed, especially with respect to socioeconomic characteristics.
Next, it was felt to be important to have both longer baseline and intervention

periods, in order to be able to control for chance factors and to simulate a more realistic situation. In view of the obvious fact that household energy consumption is influenced by a number of nonbehavioral factors (e.g. outside temperature), serious attention will be paid to how to correct for these factors.

In order to analyze possible after-effects of the experimental manipulations, post-experimental consumption data will be gathered.

Finally, given the observed tendency in mainstream behavioral energy experiments (i.e. as conducted by applied behavior analysts) to neglect cognitive variables, subjects will be extensively interviewed before and after the experimental interventions (e.g. with respect to energy attitudes and behaviors, energy conservation intentions, energy knowledge, evaluations of experimental stimuli) to trace the importance of these variables.

4.3 Research hypotheses

This section summarizes the main research hypotheses which guided this study. These hypotheses are either derived from our theoretical framework (Chapter 2) or from the review of behavioral experimentation on consumer energy conservation (Chapter 3).

The hypotheses to be tested in this field experimental study are centered around the following themes: general and specific energy attitudes (Section 4.3.1), energy knowledge (Section 4.3.2), effectiveness of information, feedback, and self-monitoring (Section 4.3.3), and some specific aspects of the functions of information, feedback, and self-monitoring (Section 4.3.4).

4.3.1 Specific and general energy attitudes

In Section 2.3.2 the position was defended that from a theoretical point of view there is convincing evidence that specific attitudes are better predictors of behavior than general attitudes. The methodological necessity of corresponding levels of specificity between attitudinal and behavioral measures was underlined. A number of hypotheses can be derived from this methodological necessity with respect to the relationship between energy attitudes and energy behaviors.

In this study (See Chapter 6) the Fishbein attitude-behavior model (Ajzen & Fishbein, 1980; Fishbein & Ajzen, 1975) will be used since this model is explicitly based on this methodological principle. According to this model, the intention to perform a given behavior is a function of the attitude toward that behavior and a normative component which consists of beliefs about normative expectations with respect to the behavior and the motivation to comply with these expectations. The attitudinal component is, in turn, a function of salient beliefs about the consequences of the behavior and evaluation of these consequences. Chapter 6 will discuss this model in more detail.

The following three hypotheses have been formulated regarding the predictive value of specific and general energy attitudes (the predictor variables are measured before the heating season starts):

H_1: specific attitudes of consumers toward energy conservation are stronger predictors of their intention to conserve energy than consumers' general attitudes toward energy scarcity.

H_2: specific attitudes of consumers toward energy conservation are stronger predictors than consumers' general attitudes toward energy scarcity of their actual energy consumption.

H_3: specific attitudes of consumers toward energy conservation are stronger predictors than consumers' general attitudes toward energy scarcity of their involvement in specific energy conservation behaviors.

Although, behavioral intentions are assumed to be immediate antecendents of behaviors, the relationship between intention and behavior depends on correspondence with respect to action, target, context, and time. Given the time gap between the different phases of this study (pre-experimental interviews, implementation period of experimental conditions, post-experimental interviews) and relevant data to be obtained in each phase, there will be considerable variation in these four factors. In view of these considerations, the following two hypotheses are formulated:

H_4: the intention of consumers to conserve energy is a moderate predictor of their actual energy consumption.

H_5: the intention of consumers to conserve energy is a moderate predictor of their involvement in specific energy conserving behaviors.

Because household energy consumption is largely an anonymous process, and thereby hardly subject to external social control, it can be hypothesized that:

H_6: attitudinal factors are stronger determinants of consumers' intention to conserve energy than normative factors.

In Chapter 2 a number of theories was discussed indicating that persons tend to favor short-term positive individual consequences of behavioral choices over long-term negative social consequences. Based on these theories the following hypothesis will be tested:

H_7: consumers' beliefs and evaluations of personal consequences of energy conservation are more determining of their intention to conserve energy, than their beliefs and evaluations of social consequences.

4.3.2 Energy knowledge

As reviewed in Section 2.3.3, behavioral energy studies have quite consistently shown that energy knowledge is hardly related to energy attitudes and behaviors. The concept of "energy illiteracy" (Ellis & Gaskell, 1978) was introduced to describe this phenomenon.
In order to validate these findings, the following hypothesis can be formulated:

H_8: there is no significant relationship between consumers' knowledge of residential energy matters and their energy consumption, their involvement in energy conserving behaviors, their intentions to conserve energy, their general attitudes toward energy scarcity, and their specific attitudes toward energy conservation.

4.3.3 Effectiveness of information, feedback and self-monitoring

Based on the theoretical discussion of assumptions underlying antecedent and consequence interventions and the conclusions drawn from the review of behavioral energy conservation experiments, the following hypotheses with respect to the effectiveness of information, feedback, and self-monitoring have been formulated.

H_9: providing consumers with information about simple and effective ways of energy conservation, as well as about personal monetary savings to be expected, results in a moderate reduction of their energy consumption.

Although most of the studies reviewed, concluded that information alone is generally quite ineffective in changing consumer energy behavior, this hypothe-

sis nevertheless assumes a positive effect. This is largely because of the fact that the information procedure used in this experiment (See Section 5.3.1) differs in some respects from the procedures applied in the studies which were discussed. It has to be added, however, that as indicated by the hypothesis, only a moderate consumption reduction is expected.
The discussion of feedback and self-monitoring yielded the following three hypotheses:

H_{10}: providing consumers with biweekly feedback on their energy consumption (amount of energy consumed, percentage increase above or decrease below baseline level, monetary consequences of increased or decreased consumption) results in a reduction of their energy consumption.

H_{11}: providing consumers with monthly feedback on their energy consumption (amount of energy consumed, percentage increase above or decrease below baseline level, monetary consequences of increased or decreased consumption) results in a reduction of their energy consumption.

H_{12}: regular self-monitoring by consumers of their energy consumption leads to a reduction of their energy consumption.

In terms of relative effectiveness of these behavioral interventions the following hypothesis is offered:

H_{13}: energy conservation information is least effective and regular self-monitoring is most effective in reducing consumers' energy consumption; in between: biweekly feedback is more effective than monthly feedback.

The fact that self-monitoring is hypothesized to be more effective than feedback - contrary to the evidence suggested by the studies reviewed - is related to the relatively long feedback intervals (biweekly, monthly) used in this experiment, compared to daily feedback frequencies often employed in those studies. The theoretical assumption behind the predicted greater effectiveness of biweekly feedback over monthly feedback, is that the formerly distinguished motivational and learning functions of energy consumption feedback (See Section 3.2.3) are likely to be more salient with biweekly feedback.

4.3.4 Specific aspects of information, feedback, and self-monitoring

As outlined in Section 3.3.3, there exists a clear tendency in mainstream behavioral experimentation on consumer energy conservation to neglect the possible influence (directly or indirectly) of cognitive and evaluative variables.

One of the conclusions was that it is hardly known in which way the responsiveness to experimental interventions is related to subjects' pre-experimental attitudes toward household energy conservation, their need for the information provided by the experimental conditions which subjects were assigned to, and their knowledge of residential energy matters. Also, little knowledge is available on the question of whether the experimental interventions evoked energy attitude change and increased knowledge of residential energy matters, or about subjects' evaluations of the conditions they were assigned to. The absence of such highly relevant information was said to seriously limit our understanding of cognitive processes involved in subjects' responsiveness to behavioral interventions. In this section some hypotheses will be offered with respect to these processes.

It could be argued that behavioral interventions to be tested in this study will be more effective when subjects have positive attitudes toward household energy conservation, have a clear need for the information provided by these interventions, and have accurate knowledge of residential energy matters. This could be called the "breeding ground theory", indicating that the effectiveness of behavioral interventions will be greater with a more favorable psychological climate or subjective breeding ground for those interventions. Assuming the validity of this theory, the following four hypotheses are formulated: (2)

H_{14}: the more positive consumers' pre-experimental attitudes toward energy conservation, the more effective information, feedback, and self-monitoring are in reducing their energy consumption.

H_{15}: the more intense consumers' pre-experimental need for energy conservation information, the more effective this information is in reducing their energy consumption.

H_{16}: the more intense consumers' pre-experimental need for regular feedback on their energy consumption, the more effective feedback is in reducing their energy consumption.

H_{17}: the more accurate consumers' pre-experimental knowledge of residential energy matters, the more effective information, feedback, and self-monitoring are as behavioral interventions promoting energy conservation.

As mentioned previously, another neglected theme in behavioral experiments on energy conservation is the relationship between induced energy conserving acts as a result of being exposed to the experimental interventions and consumers' postexperimental attitude toward energy conservation. In other words, does energy behavior change lead to energy attitude change? In terms of Bem's (1967, 1972) self-perception theory one could expect a positive answer to this question. A crucial aspect of Bem's theory is that persons evaluate their own behavior as do outside observers of that behavior. If persons behave toward some object in a given way, they will infer or attribute to themselves an attitude that is consistent with their behavior. Thus, it might be that if subjects engage in energy conservation behavior, a positive self-perception will be inferred from this behavior which finds its expression in a positive attitude toward energy conservation, which in turn may lead to persistence of conservation behavior.

H_{18}: if consumers reduce their energy consumption as a result of being exposed to the experimental interventions (information, feedback, self-monitoring), their attitudes toward energy conservation will change accordingly by becoming more positive.

It is not only assumed that subjects' self-perception will change but it is also hypothesized that their "energy literacy" will improve:

H_{19}: if consumers reduce their energy consumption as a result of being exposed to the experimental interventions (information, feedback, self-monitoring), their knowledge of residential energy matters will change accordingly by becoming more accurate.

From a policy point of view it is crucial to know the responsiveness to consumer conservation instruments by different demographic segments within the consumer population. Segmented responsiveness to these instruments implies that in order to increase consumer responsiveness the instruments themselves have to be segmented along certain demographic parameters. Though no explicit hypothesis will be offered with respect to the influence of these parameters, the data analysis will pay attention to the possibly mediating role of demographic variables.

This section will be concluded by formulating two hypotheses with regards to a) the relationship between energy conservation and comfort experience, and b) subjects' attitudes toward wider implementation of the behavioral interventions.

Winett, Hatcher, Fort, Leckliter, Love, Riley and Fishback (1982) (See also Winett, Leckliter, Love, Chinn & Stahl, 1983) observed in their experimental field study on modeling and residential energy conservation (See Section

3.3.1.2) that consumers making substantial changes in their thermostat settings reported no loss of comfort. A mean home temperature of 62°F was found, which is significantly lower that ASHRAE (American Society of Heating, Refrigerating, Air Conditioning Engineers) standards. This observed temperature, recorded by hygrothermographs, is also significantly lower than ideal comfort temperatures found in psychophysiological studies (cf. Rohles, 1981) which would predict comfort at about 75°F given the clothing worn by subjects in this experiment. "The temperature changes achieved with no loss of comfort have energy policy implications. The thermal results seriously question those who depict changes in the thermostat settings and other comparable changes as an unacceptable curtailment and "sacrificial" approach to energy conservation" (Geller, Winett & Everett, 1982, p. 172).

Based on these findings the following hypothesis will be tested:

H_{20}: consumers who reduced their energy consumption during the experimental conditions will not differ in experienced thermal comfort from consumers who did not reduce their energy consumption.

Finally, a relationship is assumed between subjects' evaluations of the behavioral interventions and their attitudes toward wider implementation of these interventions:

H_{21}: the more positive consumers' evaluations of energy conservation information, feedback, or self-monitoring, the more positive their attitudes toward wider implementation of conservation information, feedback, or self-monitoring by public utility companies.

Notes

1. This does not mean, of course, that monetary incentives in general show a lack of policy relevance. Tax rebates or subsidies for home insulation have proven to be quite effective.

2. Due to a questionnaire error no hypothesis could be tested about the relationship between subjects' pre-experimental need for regular self-monitoring and the effectiveness of self-monitoring as an experimental intervention.

PART II RESEARCH DESIGN

5. EXPERIMENTAL DESIGN

5.1 Introduction

This chapter describes the experimental design of this study in more detail. First, criteria used for selecting research locations and subjects are discussed (Section 5.2). Next, the experimental procedures are outlined, including nature of behavioral interventions, composition of experimental groups, group assignment, baseline calculations, and correction factors (Section 5.3). The following two sections (Section 5.4 and 5.5, respectively) contain an analysis of response and nonresponse regarding the pre- and post-experimental interviews with subjects. Next, the response to the post-experimental meter readings will be described (Section 5.6) and an overview will be presented of the number of subjects participating in each phase of the study (Section 5.7). After that, key-concepts in both questionnaires will be described (Section 5.8), and finally, a graphic description of the main research variables will be presented (Section 5.9).

5.2 Research settings and subjects

The experiment was conducted in five towns in the Netherlands from January to July 1981. Subjects were approximately 400 housewives recruited from five neighborhoods of physically identical, although with different wind orientations, mainly single-family dwellings with individually metered gas-fired central heating systems. Participants were randomly assigned to either the experimental groups or the control group and have been extensively interviewed before and after the experiment.

The experimental target group are housewives. The rationale of choosing housewives as subjects for this study is simply the assumption that their behavior during the day (e.g. thermonstat setting, airing, washing) probably is one of the most important nontechnical factors explaining household energy consumption. This assumption, of course, is only valid for traditional sex role patterns in households. An additional motive for choosing housewives as experimental subjects is the advantage of a somewhat homogeneous target group which facilitates the design of our behavioral interventions. As such it is possible to tailor these interventions to existing attitudes, opinions, and behaviors of the target group, which is a central principle of communications theory (van den Ban, 1982; van Woerkum, 1982) and social marketing.

Because of practical reasons research settings have been selected within the boundaries of the province of North-Holland, the province in which the Free University is located. In selecting eligible neighborhoods the following criteria were applied:

1. in order to be able to attribute differences in experimental outcomes to differences in experimental conditions it is necessary in this study to especially control for dwelling characteristics. Therefore, the first criterion was to select neighborhoods with physically identical houses.

2. in several stages of the experiment subjects could - at least theoretically - terminate their co-operation, therefore the neighborhoods had to have a sufficient number of similar houses to result in a minimum of 60 subjects per condition having completed all experimental phases. For this reason neighborhoods were looked for with no less than 200 physically similar houses.

3. the penetration grade of gas-fired central heating systems in the residential sector of the Netherlands is more than 60%, almost every new house is centrally heated. In combination with the fact that 70% of Dutch houses are single family-dwellings, a next criterion - thereby increasing the external validity of the study - was to select neighborhoods with single-family dwellings provided with gas-fired central heating systems.

4. in order to be able to determine the effectiveness of the behavioral interventions at the household level and to make the necessary calculations for the feedback conditions each house had to be provided with an individual gas and electricity meter.

5. there had to be some spreading between neighborhoods in urban/rural differences as well as in year of construction of the houses.

In consultation with a number of municipalities and housing corporations in North-Holland the following neighborhoods from five different towns in the province have finally been selected as research locations:

1. Heerhugowaard (31.500 inhabitants) : Molenwijk
2. Zaandam (129.000 inhabitants) : Plan Kalf
3. Purmerend (32.500 inhabitants) : Wheermolen
4. Hilversum (93.000 inhabitants) : De Meent
5. Amsterdam (717.000 inhabitants) : Osdorp

As it turns out most selection criteria could be met quite satisfactorily. There are two exceptions. First, the total number of houses in the Molenwijk location in Heerhugowaard (See table 5.1) does not exceed 200 (criterion two). Second and more seriously, no location in Amsterdam could be spotted which satisfied all criteria. This was mainly due to the fact that no sufficient number of physically indentical single-family dwellings could be found. Consequently, as far as the Amsterdam location (Osdorp) is concerned multi-family dwellings had to be selected.
Table 5.1 describes the main physical characteristics of the five research settings (see next page).

Apart from the Amsterdam location the research settings include three-bedroom single-family dwellings with living room, bathroom, nonheated attic, and with or without an open kitchen. In accordance with selection criterion five there is an acceptable spreading in urban/rural differences as well as in year of delivery of the houses. As a direct consequence of criterion one and two the number of people owning their home in the research locations is strongly underrepresented. The Heerhugowaard location is an exception to this rule as here 70% of the occupants are home owners.
As can be seen from Table 5.1 the research settings differ markedly in insulation provisions. The Amsterdam and Hilversum location lack any form of insulation, whereas the Heerhugowaard dwellings are fully insulated in line with current insulation norms for new buildings. Recently, most houses in the Zaandam and Purmerend location have been provided with wall insulation and double glazing in the living room. For a correct understanding of Table 5.1 it has to be added that the data are based on information by housing corporations. As such these data do not reflect possible changes in the homes made by occupants themselves; these changes will be traced through the pre-experimental interviews with subjects.
There are some differences in heating systems between locations. Houses in the Purmerend neighborhood are equiped with an one-pipe central heating system with convector, whereas the multi-family apartment dwellings in the Amsterdam location have central heating stoves. Houses in the other three research settings are equiped with standard central heating systems: two-pipe systems with radiators and separate furnace.
Finally, the houses in the Zaandam neighborhood are provided with an electric boiler.

In this section a short description of some physical characteristics of the five research settings has been presented. In Section 5.4 the selection of sub-

Table 5.1: Some Physical Characteristics of the Five Research Settings

	Heerhugo-waard	Zaandam	Purmer-end	Hilver-sum	Amster-dam
- type of dwelling	single-family	single-family	single-family	single-family	multi-family
- year of construction	1979	1975	1967/68	1975	1960
- total number of houses	174	219	240	200	216
- percentage of rented houses	30%	100%	99%	100%	100%
- number of rooms	4	4	4	4	2/3/4
- floor surface[1]	46-51 m^2	47 m^2	42 m^2	50-52 m^2	56/62/70 m^2
- percentage of houses with open kitchen	47%	100%	-	100%	-
- percentage of houses with electric boiler	-	100%	-	-	-
- percentage of attic insulation	100%	-	-	-	-
- percentage of floor insulation	100%	-	-	-	-
- percentage of wall insulation	100%	52%[2]	84%[3]	-	-
- double glazing in living room	100%	88%[2]	84%[3]	-	-

[1]) outwork measured ground floor for single-family dwellings; total floor for multi-family dwellings
[2]) according to the 1980 planning
[3]) according to the 1977 returns

jects from these locations will be discussed in more detail. First of all, however, the experimental conditions and procedures will be outlined.

5.3 Experimental design and intervention strategies

As has been described in the preceding chapter, three intervention strategies have been chosen to be tested experimentally: energy conservation information, feedback on household energy consumption (biweekly or monthly), and self-monitoring by subjects of their own energy consumption.

These three intervention strategies yielded four experimental groups and one control group:

1. Information-only Group: subjects received a booklet - specially written and designed for this experiment - with practical household energy conservation tips. As housewives are the target group of this study the conservation information focuses on relevant energy aspects of the chronological household activities of housewives during the day and suggests numerous conservation possibilities.
With the theoretical framework as outlined in Chapter 2 in mind (particularly Olson's theory of the logic of collective action) a salient aspect of the information was individual financial savings from energy conservation.

2. Biweekly Feedback Group: in addition to receiving the energy conservation booklet, subjects were informed every two weeks about their gas and electricity consumption in the previous two weeks, the percentage increase or decrease compared to their 1980 baseline consumption, as well as an estimate of the financial consequences of these changes in their energy consumption projected on a monthly basis.

3. Monthly Feedback Group: received essentially the same information as the biweekly feedback group but on a monthly basis.

4. Self-monitoring Group: subjects were asked to read their (inside) gas and electricity meter regularly and to record their energy consumption on an uniform recording chart. These recording charts instruct participants how to calculate consumption differences and how to compute monetary consequences of increased or decreased energy consumption. As in the other experimental groups, subjects in this condition also received the energy conservation booklet.

5. Control Group: apart from being interviewed before and after the period the experimental conditions were in effect, no special treatment was given to this control group.

In order to assign the 400 housewives to one of these five groups the following random matching procedure was used. Prior to the experimental period (January to July 1981) all selected subjects (see Section 5.4) were extensively interviewed in October and November 1980 on energy and energy-related matters. Without any reference to the experiment, subjects were asked to give their (written) permission to the research team to obtain their last year's natural gas and electricity consumption figures from the utility companies. By converting gas and electricity consumption figures to joules into one energy consumption score, subjects, giving their permission could be ordered per location according to their consumption scores. Using random numbers each group of five successive subjects was assigned systematically to the five conditions. When executing this matching procedure complete data were only available for subjects from the Amsterdam and Zaandam locations. Participants from Heerhugowaard and Purmerend were only matched for gas consumption, whereas in Hilversum due to administrative problems subjects were matched according to physical heating needs of their houses. The results of this matching procedure will be discussed in Section 5.4.

The experimental design is a simple ABA order (Baseline - Treatment - Baseline) of experimental conditions with a no-treatment control group. In order to test the hypotheses about possible long-term behavioral effects of these conditions, meter readings of those subjects having completed the experiment were recorded about six months (in January 1982) after the experimental conditions were withdrawn.

In the next three subsections the practical course of affairs with regards to each behavioral intervention will be further specified.

5.3.1 Information

One of the main advantages of energy conservation information as a communicative change strategy is obviously the combination of large-scale applicability and relatively low costs.

Based on existing information sources a manual on energy conservation procedures was compiled in close cooperation with the Dutch Foundation for Energy Conservation Information (SVEN) and the Dutch Department of Economic Affairs. Given the design of this study, housewives are the target group of the manual. It was decided that the conservation information should be presented along the following lines:

1. all relevant conservation tips with regards to both natural gas and electricity consumption have to be collected in one manual.

2. the conservation tips should focus on concrete and practical behavioral changes.

3. financial savings from energy conservation must be the main message of the

manual.

4. the conservation tips have to be arranged in such a way that they fit a cognitive framework which is familiar to the target group.

5. special attention should be given to the readability and comprehensibility of the manual.

With these general guidelines in mind a manual or booklet was compiled consisting of nine solid cardboard charts with each chart focusing on a particular energy conservation theme. The format of the manual is 30 by 21 cm, a format which in combination with its attractive multi-colored design enables subjects to pin the charts - kept together by a plastic clasp - to the wall. In line with our theoretical considerations the running head of the manual is "How much money you can save on your natural gas and electricity consumption".
The energy conservation tips are ordered in such a way that they follow the chronological order of normal household activities during the day, e.g. getting up in the morning, day-time thermostat setting, doing the washing, preparing meals, lighting the home in the evening, night-time thermostat setting. Within the domain of each of these everyday activities specific energy conservations tips are discussed and suggested. Estimates were given of the financial savings from these conservation procedures. In addition to these procedures the charts contain information about energy use of appliances, energy billing, and about how to record one's own household energy consumption and consumption changes through regular meter readings.
By structuring the conservation information in this way an effort was made to match the housewives' frame of reference.
Mid-January 1981 the energy conservation manual was mailed to the subjects.

5.3.2 Feedback

As has been outlined in previous chapters, practical and policy considerations guided the final choice for the two feedback frequencies (every two weeks, every four weeks) evaluated in this study. This section describes the feedback procedure.

Meter readings (both natural gas and electricity) were recorded by local or regional utility company employees on Mondays at about the same time and in about the same sequence by the same employee. If no adult was at home the household was contacted in the evening by telephone by the research team and

one of the adult household members was asked to read the meters.
The meter readings were computerized and after some statistical corrections (see below) for outside temperature (gas) and seasonal fluctuations (electricity) standard feedback forms (30 cm by 21 cm) were mailed to subjects in both feedback conditions on the Tuesday and Wednesday after the meter reading. The feedback forms provided the following information:

- date of meter reading
- gas and electricity consumption in the preceding two or four weeks
- the percentage increase or decrease (after correction) in relation to subjects' 1980 level of energy consumption
- the amount of money corresponding with this increase or decrease projected on a monthly basis

In total the biweekly group received feedback 12 times (from January 6 to June 22, 1981) and the monthly feedback group 6 times.
Hardly any household was missed in both conditions. There are two exceptions to this rule. The first time feedback was given, subjects from the Hilversum location had to be excluded because their 1980 baseline consumption was not known yet. Some subjects from the Amsterdam location did not receive their feedback the first time because their meter readings were not available at that time.

Subjects in the two feedback conditions were contacted by letter in January 1981. The feedback procedure was explained (including the goal of the experiment, the regular visit of the utility company employee for the meter readings, the feedback information, and the duration of the experiment). Together with this letter subjects received the energy conservation manual.

The next two subsections discuss the correction factors that have been used to eliminate certain influences that may confound the feedback information. These influences are related to the effects of changes in outside temperature on natural gas consumption and to certain seasonal effects on electricity use.

5.3.2.1 Correction factors natural gas

As - among other things - a direct effect of changes in outside temperature, household gas consumption is unevenly spread over the year. Therefore, in order to compare gas consumption in the experimental period with baseline consumption (the essence of energy feedback) one has to correct for outside temperature. In this experiment the degree day method was used. In short, this method converts baseline gas consumption into outside temperature units. Next, the total number of outside temperature units in the experimental feedback period is multiplied by gas consumption per temperature units during baseline consumption. The outcome of this computation is a predicted level of gas consumption during the experimental period corrected for outside temperature differences between baseline and experimental period. Other correction methods (e.g. individual household regression method, control group method) have been considered but were not applied because of the fact that from a policy point of view their use on a large scale is infeasible, at least as provided by the studies reviewed in Chapter 3.

Subjects' 1980 gas consumption was used as their baseline consumption. From this baseline their experimental use had to be predicted. This prediction was calculated as follows. The mean temperature per day at which one does just not heat the house is called the "reference temperature". For each day during the baseline period the number of degrees of the mean temperature below the reference temperature is counted and summed over the period concerned. By dividing total gas consumption in this period by the total number of computed degree days an indicator is developed for mean gas consumption per degree day. During the experimental feedback period the total number of degree days for each sequence of two or four weeks (depending on the feedback condition) was counted and multiplied by mean degree day baseline gas consumption. The following feedback formula was then used:

$$\frac{PC - EC}{PC} \times 100$$

where:

PC = Predicted Consumption

EC = Experimental Consumption

As a certain part of household gas consumption does not vary with outside temperature (e.g. cooking, hot water use) predicted gas consumption had to be raised with an estimate of this relatively nonvariable, fixed consumption. This estimate was set at 500 cubic meters per year, which as such is a current estimate. For the Zaandam location (houses with an electric boiler) this estimate was set at 250 cubic meters per year.

The reference temperature was fixed at 15° C for noninsulated houses and 14° C for houses with at least wall insulation or double glazing in the living room. To determine mean outside temperatures both during baseline and experimental periods, official readings from regional weather stations were used.

Apart from these correction factors differences between cornerhouses and interjacent houses and between insulated and noninsulated houses were eliminated by means of a weighting procedure.

The degree day method certainly has its limitations (see Meyer, 1981). Sun and wind influences are not directly accounted for and especially in extremely cold or mild periods the degree day method may lead to inaccuracies. Over longer periods, however, the method gives a reasonable picture of increasing or decreasing tendencies in gas consumption. Also, it is a fairly practical method which is not true for alternative methods like individual regression methods or control group methods. (1)

5.3.2.2 Correction factors electricity

In houses with gas-fired central heating systems such as in these research locations electricity consumption shows a decreasing tendency in the January-July period. This tendency is largely a consequence of the increasing amount of daylight in this period. This means that electricity consumption is not proportionally spread over the weeks within the year. For providing feedback on electricity consumption and for establishing the effectiveness of the other behavioral interventions this implies that one has to correct for this disproportional spreading. To determine the required strength of this correction factor a secondary analysis was done on a data file from the Provincial Electricity Company North-Holland (P.E.N.) which listed weekly electricity consumption for the year 1977 of 1464 houses from a medium-sized town (Den Helder) in North-Holland. For each week number the cumulative proportion of the yearly electricity consumption was calculated. Next, a regression analysis was done of these cumulative proportions on the week number. The regression outcome served as a basis to draw a reasonably fitting curve for the first 26 weeks of the year.
Based on this curve for each successive interval of two to four weeks (feedback conditions) and for the experimental period as such (other conditions) the expected percentage of the yearly electricity consumption could be computed.
As subjects from the Zaandam location have electric boilers, adjusted percentages for this location had to be computed. It was estimated that approximately one third of their electricity consumption was proportionally spread over all weeks of the year. Consequently, the season correction factor is related to two thirds of their electricity consumption.(2)

5.3.3 Self-monitoring

Active self-feedback on household energy consumption and on energy consumption changes is the main rationale behind the self-monitoring condition. More specifically, the basic goals of this condition are to find out whether consumers are willing and able to monitor their own energy use, to see how they handle such a relatively simple behavioral technique (e.g. frequency chosen, degree of recording completeness), and to determine whether conservation acts occur as a consequence of self-monitoring.

The self-monitoring meter reading recording form used in this experiment was an existing recording form devised by the Dutch Foundation for Energy Conservation Information (SVEN). This form is a nine-column solid cardboard chart of 29.5 by 21 cm with the following motto (translated) as headline: "Be wise with energy, keep the score".

Information that could be recorded in the columns includes: recording date, number of days relative to the prior reading, meter reading of electricity (KWH) and gas (cubic meters), absolute electricity and gas consumption relative to the preceding reading, and mean electricity and gas consumption per day since last reading. Also, space was provided to record mean outside temperatures and to record special circumstances. In order to increase the salience of financial consequences, electricity and natural gas prices - respectively per KWH and per cubic meter - were listed.

Finally, to instruct subjects how to make the appropriate calculations a number of examples were printed.

With daily meter readings the number of lines on the form was sufficient for six weeks. During the experimental period subjects in the self-monitoring condition received a recording form three times (mid-January, early March, and late April 1981).

In the letter accompanying the first form the usefulness of regular meter reading was outlined, and subjects were asked to record their meter readings at least once a week and were instructed how to use the recording form. Subjects were told that they would receive a new recording form each six weeks until the summer together with a return envelope. This envelope was to be used to return the completed form to the research team. On receipt of these completed forms, photocopies were made of the meter readings and calculations subjects had made, and next the original forms were returned to the participants. This last procedure enabled subjects to make cross-period comparisons.

In the second (March) and third (April) letter subjects were prompted to record their meter readings daily, and to relate the readings to mean daily outside temperature. Together with the second letter subjects also received a plastic SVEN color indicator which turns different colors with different inside temperatures. As such, the indicator gives rough but continuous visible inside tem-

perature estimates.
Subjects in the self-monitoring condition also received the energy conservation manual.

5.4 Recruitment of subjects and pre-experimental interviews

This section describes the way subjects were recruited from the five selected research locations and their response to the baseline interviews preceding the experimental period. In the following subsections a demographic profile of our subjects will be sketched both with respect to research location and with regard to experimental condition to which subjects were assigned. This last analysis also serves as a check on the earlier described matching procedure.

5.4.1 Recruitment of subjects

The following guidelines directed the recruitment procedure for this study. Starting-point was the principle that about 300 subjects had to complete all phases of the experiment (i.e. baseline interview, permission to obtain 1980 energy consumption data, exposure to experimental conditions, and postexperimental interview); proportionally this means 60 subjects per location and per condition. Mortality during and after the experimental phase was estimated at 25%, the nonresponse to the pre-experimental (baseline) interviews at approximately one-third. These rough estimates implied that prior to the baseline interviews about 600 (600 x .75 x .67 = 300) subjects had to be recruited. As will be shown in the next subsection the estimate of nonresponse to the baseline interviews was too optimistic, and an additional sample of addresses became necessary.
Before the actual interview took place, an introductory letter from the Institute for Environmental Studies, Free University was mailed to each address. In order to avoid a systematic response bias with regard to the topic of this study, no explicit reference was made in the letter to energy conservation issues. The study was simply introduced as one in a series of studies this institute conducts on environmental issues.
Subjects were interviewed in September and October 1980 by skilled interviewers from a professional public opinion research institute (Veldkamp Marktonderzoek BV, Amsterdam). In all, some 80 interviewers participated after being extensively instructed about the study. The questionnaire was fully structured.

5.4.2 Response

As already indicated, the initial nonresponse turned out to be much higher than the original estimate. Therefore, a second interview round with additional sample addresses was decided upon. Almost 300 extra households were approached for an interview. As table 5.2 shows this decision resulted in a final response of 55%.

Table 5.2: Response to Pre-Experimental Interviews

	First Round	Second Round	Total
Number of households approached	574	305	879
- no housewife in household	12	3	15
- no fluency in Dutch	7	7	14
final response	555	295	850
	(56%)	(54%)	(55%)

However, besides the absolute response it is the relative response, i.e. the response between research locations, which is especially important to this study. Table 5.3 provides more information about response differences between the five locations.

Table 5.3: Response to Pre-Experimental Interviews by Research Location

Location	Total Number of Households Approached	Absolute Response	%
Heerhugowaard	129	94	73%
Zaandam	168	105	62%
Purmerend	177	87	49%
Hilversum	180	91	51%
Amsterdam	196	93	47%
Total	850	470	55%

Obviously, there are marked response differences between the research locations. It turns out that Heerhugowaard, with a response of 73% is the only location meeting the original response estimate. Zaandam (62%) appears to have the second highest response, whereas the Amsterdam location yields the lowest response (47%). One has to add that as far as this last city is concerned, market research experience shows that this response rate is even quite high for Amsterdam.

An inventory has been made of causes accounting for the nonresponse phenomenon in this phase of the study. Table 5.4 summarizes the main causes by research location.

Table 5.4: Nonresponse Cause by Research Location

Causes	Heerhugowaard	Zaandam	Purmerend	Hilversum	Amsterdam	Total
Not home after three trials	23%	17%	6%	19%	23%	17%
Mentally or physically unable	3%	6%	9%	9%	10%	8%
Husband interviewed by mistake	11%	-	-	-	1%	1%
Refusal	63%	76%	85%	81%	66%	73%)
	(N=35)	(N=63)	(N=89)	(N=90)	(N=103)	(N=380

As table 5.4 indicates, refusal to agree with an interview is the single most important nonresponse cause in all locations. There is some variation in this respect between locations in the sense that refusal is a relatively less important nonresponse cause in Heerhugowaard and Amsterdam. The second most important cause is the absence of housewives after three contact trials. Subjects being mentally or physically unable to be interviewed account for 8% of the total nonresponse.
Finally, the percentage of subjects (men) who were interviewed by mistake in Heerhugowaard is remarkably high.

If one looks at the reasons subjects gave to refuse an interview it appears that 75% said they had no time or were not interested in the topic of the study.
One could hypothesize that the fact that both goal and content could not be stated in explicit terms - in order to avoid response selectiveness - decreased

the willingness to cooperate.
The interviews themselves were checked by randomly sending evaluation forms to a subsample of 231 subjects. One hundred forms were returned, yielding a response rate of 43%. It turns out that 91% of the respondents indicated that they thought the study was useful, and 83% that the interview was pleasant and agreeable. No irregularities were reported.

In the introductory letter subjects were informed that the interview would take less than one hour. This appeared to be a rather accurate estimate as the mean interview length was 60 to 65 minutes; nevertheless some 20% of the interviews took more than 75 minutes. In the evaluation forms no complaints about the length of the interview were reported.

5.4.3 Willingness to agree to a second interview and permission to obtain energy consumption rates

As described in section 5.2 the experimental design made it necessary that without any explicit reference to the experimental follow-up subjects had to be willing to agree to a second (post-experimental) interview and had to give their permission to obtain their 1980 energy consumption rate from the utility company.
Subjects' response to the first request is summarized in Table 5.5

Table 5.5: Willingness to Agree to a Second Interview

Location	Willing	Not Willing	Otherwise
Heerhugowaard	88%	1%	11%
Zaandam	95%	-	5%
Purmerend	95%	3%	2%
Hilversum	87%	-	13%
Amsterdam	94%	3%	3%
Total	92%	1%	7%

It is certainly gratifying to observe that on an average 92% of the respondents agreed with a second interview. There are some differences according to location. The announcement that the second interview would be considerably shorter than the first one has probably influenced the response in a positive way.

Having subjects' last annual energy consumption is of crucial importance to this study in order to assess a reliable baseline level. Therefore, subjects were asked to sign a paper permitting the research team to obtain their most recent (1980) annual consumption of natural gas and electricity from the utility company files.

Subjects who preferred first to consult their spouse or other household members, received a form which they could fill out and return later. Because in a number (32) of cases it was unclear whether such a form was actually left or not, a letter was sent asking subjects to give their written permission.

Table 5.6: Permission for Obtaining Subjects' 1980* Energy Consumption Rate

	Heerhugowaard	Zaandam	Purmerend	Hilversum	Amsterdam	Total
Permission:						
- in questionnaire	73%	82%	79%	88%	92%	83%
- via form	3%	3%	5%	-	-	2%
- via letter	4%	2%	1%	5%	1%	3%
No permission	5%	4%	4%	3%	6%	5%
No reaction	14%	10%	10%	3%	-	7%
Both permission and agreement to second interview	81% (N=76)	87% (N=91)	85% (N=77)	93% (N=81)	92% (N=86)	87% (N=411)

* By 1980 energy consumption rate (baseline consumption) we mean subjects' most recent annual consumption rate the utility company disposed of prior to the start of the experiment (January, 1981). This annual rate does not necessarily cover the period from January 1 to December 31, 1980.

Table 5.6 shows that only 5 percent of all subjects refused to give their permission. The overwhelming majority agreed with the request to have their consent for obtaining their consumption data. Remembering the former rule that 80 subjects per location were needed prior to the start of the experimental treatment who both agree with a second interview and give their permission for obtaining their 1980 energy consumption data, one may conclude that the objective has been realized quite satisfactorily.

A large number of analyses was done - using variables from the pre-experimental survey - to see whether subjects not giving their permission to either obtai-

ning their 1980 energy consumption rate or to the second interview also differed in other aspects. It turned out that this was only marginally the case. They appeared to be slightly less knowledgeable about how much they pay every two months for gas (R = -.12, N = 441) and have somewhat less positive evaluation of conservation on their gas and electricity consumption (R = -.14, N = 470).
Of the 411 subjects mentioned in Table 5.6, 406 were finally included in the matching procedure.

5.4.4 A demographic profile of subjects

Previous research (See Chapter 2) has indicated that household energy consumption is related to a number of demographic variables. Therefore, it is important to know the demographic composition of this sample. This subsection will sketch a demographic profile of subjects according to research location. The following variables will be included: family life cycle, household size, socioeconomic status of household, housewives' age, education, occupational time spending, and age of children. Data will be provided which enable a comparison with national figures.
The data set pertains to subjects (N = 470) who completed the pre-experimental interview.

Table 5.7 (see next page) shows that the majority of sample households are married couples with children younger than 18 years. Heerhugowaard has a high proportion of two-person households in which the housewife is under 35 years, whereas in Purmerend a relatively high percentage is found of households without children in which the housewife is over 35 years. Zaandam and Hilversum show a fairly similar family life cycle pattern.
As far as household size is concerned, it appears that Heerhugowaard - the most recently built location - has a relatively high proportion of two-person households. Again, Zaandam and Hilversum show a similar composition in terms of number of persons per household. The Purmerend location appears to be quite comparable with the national picture.

With respect to subjects' age, Table 5.8 (see page 100) indicates that on an average housewives are younger compared to the national figures. Amsterdam shows the greatest variation in age, whereas over 75% of housewives from the Heerhugowaard location is under 35 years. Proportionally, Purmerend has a high percentage of subjects over 50.
Mean age distribution of subjects' children within the total sample resembles

Table 5.7: Family Life Cycle and Family Size of Subjects by Research Location

	Heerhu-waard	Zaan-dam	Purmer-end	Hilver-sum	Amster-dam	Total	The Nether-lands*
N =	94	105	91	87	93	470	9580
<u>Family life cycle</u>							
Single $<$ 35 years	-	-	-	-	-	-	2%
Single $>$ 35 years	-	-	2%	-	12%	3%	10%
Married couple							
- Wife $<$ 35 years	40%	2%	-	2%	9%	11%	9%
- Wife $>$ 35 years	4%	9%	35%	7%	22%	15%	21%
Family with children $<$ 17 years	55%	90%	63%	91%	56%	71%	49%
<u>Family Size</u>							
one person	-	-	2%	-	12%	3%	11%
two persons	43%	8%	19%	7%	29%	21%	30%
three persons	21%	17%	23%	13%	20%	19%	19%
four persons	21%	55%	34%	53%	28%	38%	25%
five persons	11%	19%	14%	18%	5%	14%	10%
six persons or more	4%	1%	8%	9%	6%	6%	5%

* <u>Source</u>: Minicensus 1980, Attwood-Interact, Dongen, 1981. (Dutch households in which wife is responsible for the housekeeping).

Table 5.8: Subjects According to Age and Age of Children Living at Home Under 18 Years by Research Location

	Heerhu-waard	Zaan-dam	Purmer-end	Hilver-sum	Amster-dam	Total	The Nether-lands*
N =	94	105	91	87	93	470	9580
<u>Age</u>							
⩽ 24 years	33%	1%	-	5%	6%	9%	8%
25 - 34 years	45%	39%	11%	41%	32%	34%	27%
35 - 49 years	19%	50%	49%	47%	29%	39%	31%
50 - 64 years	3%	10%	37%	6%	14%	14%	23%
65 - 74 years	-	-	2%	-	14%	3%	8%
⩾ 75 years	-	-	-	1%	5%	1%	3%
<u>Age of children</u>							
⩽ 5 years only (A)	25%	3%	3%	9%	18%	11%	12%
6 - 12 years only (B)	4%	30%	12%	33%	9%	18%	11%
13 - 17 years only	4%	15%	30%	12%	13%	15%	13%
A + B	14%	14%	2%	17%	6%	11%	7%
A + C	-	1%	-	2%	1%	1%	-
B + C	7%	20%	14%	15%	7%	14%	9%
A + B + C	1%	-	1%	2%	2%	1%	1%
no children ⩽ 17 years	45%	10%	37%	9%	43%	29%	46%

* <u>Source</u>: Minicensus 1980, Attwood-Interact, Dongen, 1981. (Age of housewife/husband and age class of children of Dutch households).

the national distribution. It appears that households with children younger than 5 years are to be found especially in Heerhugowaard, and households with children between 6 and 12 years in Zaandam and Hilversum.

It can be observed from Table 5.9 that by comparison subjects from Heerhugowaard have the highest socioeconomic status and educational levels, and are most likely to have a paid job of more than 20 hours per week. Least differences are found between Zaandam and Hilversum. Comparing the sample with Dutch households as such, it seems that both housewives with higher and lower socioeconomic status and education are somewhat underrepresented. As far as occupational situation is concerned, Table 5.9 shows that in this sample the percentage of housewives having a paid job is higher than the national figure.

In summary, the above analyses have indicated that with respect to demographic characteristics, the demographic profile of the sample is quite heterogeneous. In view of further analyses of findings and outcomes from the experiment, this variation is quite satisfactory.

It has to be added that a subsequent analysis revealed no statistically significant differences in demographic variables between the sample of subjects (N = 470) which completed the pre-experimental interview, the sample (N = 411) that was randomly assigned to either the experimental groups or the control group (i.e. subjects who both gave their permission to obtain their 1980 energy consumption rate from the public utility company and agreed to a second interview), and the sample (N= 387) of subjects which competed all phases of the experiment.

5.4.5 Matching results

As described in Section 5.3.1 subjects were randomly assigned to either one of the fthe experimental groups or the control group. An analysis of variance (ANOVA) revealed no statistically significant ($p < .05$) differences between the five groups with respect to both corrected baseline natural gas consumption (F (4.381) = .40, p = .81) and corrected baseline electricity consumption (F (4.382) = .79, p = .53).
A chi-square test was performed to see if assigned group members differ in demographic characteristics as specified in section 5.4.4. Results indicate no statistically significant differences between the five groups with regard to neither family cycle (χ^2 = 13.80, df = 12, p = .31), household size (χ^2 = 16.85, df = 20, p = .95), housewife's age (χ^2 = 7.95, df = 20, p = .99)), edu-

Table 5.9: Subjects According to Socioeconomic Status, Education, and Occupational Time Spending by Research Location

	Heerhu-waard	Zaan-dam	Purmer-end	Hilver-sum	Amster-dam	Total	The Nether-lands*
N =	94	105	91	87	93	470	9580
Socioeconomic Status							
A/B (high)	10%	8%	9%	5%	3%	7%	12%
C_1	17%	13%	17%	13%	4%	13%	15%
C_2	20%	32%	23%	35%	15%	25%	21%
D_1	45%	43%	34%	40%	44%	41%	41%
D_2	2%	-	2%	-	8%	3%	11%
no information	6%	4%	15%	8%	25%	12%	
Education							
lower education	11%	17%	26%	10%	38%	21%	34%
lower educational plus vocational education	26%	37%	35%	36%	26%	32%	19%
continued education	18%	23%	13%	18%	18%	18%	10%
continued education plus vocational education	18%	13%	15%	26%	6%	16%	14%
secondary and university education	23%	10%	8%	9%	9%	12%	16%
no information	2%	-	1%	-	2%	1%	7%
Occupational Time Spending							
paid job of 20 hours or more	50%	18%	16%	7%	16%	21%	11%
paid job less than 20 hours	4%	21%	13%	18%	10%	14%	8%
no paid job	46%	61%	71%	75%	74%	65%	80%

* Source: Minicensus 1980, Attwood-Interact, Dongen, 1981. (Socioeconomic status of households, education of housewife/husband (all households), and paid occupation of housewife/husband responsible for the housekeeping)

cation (χ^2 = 27.74, df = 20, p = .12), occupational time spending of housewife (χ^2 = 3.03, df = 8, p = .93), nor to age of children (χ^2 = 31, df = 28, p = .29).
In summary, one may conclude given these results that the matching procedure was successful.

5.5 Response to the post-experimental survey

As outlined before, subjects were again orally interviewed in June/July 1981 after the experimental conditions had been withdrawn. As much as possible subjects were interviewed by the same interviewer who administered the pre-experimental questionnaire. Forty interviewers participated in this post-experimental survey.
The questionnaire was again fully structured. If neccessary, the same measurements were used as included in the pre-experimental survey. Moreover, a set of specific questions was asked with regard to the experimental conditions subjects were assigned to.
Table 5.10 shows the response rate both by research location and by experimental condition.

Table 5.10: Response to Post-Experimental Interviews by Research Location and by Experimental Condition

Location	Response	N (= 100%)
Heerhugowaard	95%	76
Zaandam	86%	90
Purmerend	88%	76
Hilversum	95%	81
Amsterdam	67%	83
Total	86%	406

Experimental Condition	Response	N (= 100%)
Information	84%	82
Biweekly Feedback	91%	81
Monthly Feedback	88%	81
Self-Monitoring	80%	81
Control	86%	81

Table 5.10 indicates that apart from the Amsterdam location the response to the post-experimental survey was quite high. It appears, that the main nonresponse reason in Amsterdam is absence of subjects at several successive contact attempts by the interviewer. With the exception of the self-monitoring group, no pronounced differences in response exist between experimental groups. As will be shown in later chapters the relatively higher interview refusal in the self-monitoring group is related to compliance with the self-monitoring request itself: subjects who refused the interview also tended to refuse the self-monitoring request.
Again, the interviews were checked by randomly mailing evaluation forms to a subsample of 85 subjects. Thirty subjects returned the forms, yielding a response of 30%. It turned out that 93% of the subjects expressed a positive attitude toward the study. All subjects indicated that the interview was pleasant and agreeable. No irregularities were reported.

In regards to the starting-point (See Section 5.4.1) that 60 subjects per location and per condition should complete all phases of the experiment, one may conclude that this objective has been realized.

5.6 Post-experimental consumption data

In order to discern possible long-term effects of the experimental interventions, data were gathered about subjects' post-experimental energy consumption. In January 1982, approximately six months after termination of the experimental conditions, subjects were mailed a letter asking them to read their gas and electricity meter and to record the meter readings - including recording date - on an enclosed form and to return this form. If neccessary, subjects were contacted by telephone to obtain their meter readings. Table 5.11 indicated the response to this meter reading request.

Table 5.11 (see page 105) shows a satisfactory overall response rate of 93%. Again with the exception of the Amsterdam location there are no striking differences in compliance with the meter reading request between research locations. The same conclusion applies to differences between experimental conditions.

5.7 Overview of observations in various study phases

Though the contact itself was kept impersonal, there were various points during the study at which subjects were approached.
For matters of clarity it therefore may be helpful to give a synopsis of the

number of observations obtained in the different phases of this study. Table 5.12 (see page 106) contains this synopsis.

Table 5.11: Response to the Post-Experimental Meter Reading Request by Research Location and by Experimental Condition

Location	Response	N (= 100%)
Heerhugowaard	97%	76
Zaandam	94%	90
Purmerend	95%	76
Hilversum	100%	81
Amsterdam	77%	83
Total	93%	406

Experimental Condition	Response	N (= 100%)
Information	91%	82
Biweekly feedback	95%	81
Monthly feedback	94%	81
Self-monitoring	91%	81
Control	91%	81

Table 5.12: Synopsis of Number of Observations in Various Phases of the Study

Period	Study Phase	Number of Observations n
September/ October 1980	Pre-Experimental Interviews	
	- selected addresses	879
	- valid addresses	850
	- completed interviews	470
November/ December 1980	Baseline Measurement	
	- obtained data on subjects' 1980 energy consumption	411
	- usable consumption data	406
January/ June 1981	Implementation of Experimental Procedures and Interventions	
	- subjects assigned to experimental conditions	406
	- first meter reading	400
June/July 1981	End of Experimental Period/Post-Experimental Interviews	
	- second meter reading	387
	- completed interviews	349
January 1982	Post-Experimental Consumption Data	
	- third meter reading	376

5.8 Measurement of key-concepts

The next two subsections will briefly list which key-concepts have been measured in both the pre- and post-experimental questionnaire. Subsequent chapters will provide further details on these measurements.

5.8.1 The pre-experimental questionnaire

This 33-page questionnaire was fully structured and contained questions related to the following set of variables:

- demographic variables: age, education, and socioeconomic status of both housewife and husband, household composition, household income, number of years living in present dwelling, home owner/renter.

- dwelling characteristics: heating system, hot water preparation, insulation measures, number of rooms, type of rooms, possible rebuildings. (3)

- household energy behavior: day- and night-time thermostat setting both during presence and absence, radiator use, insulation provisions, use of curtains, furnace flame and hall-door, electrical appliances use, purchase behavior, electric light use, washing and bathing habits.

- energy knowledge: price per cubic meter gas and per KWH electricity, bimonthly advance pay for gas and electricity, amount of gas and electricity consumed in the preceding billing year, knowledge of insulation subsidies, estimate of electricity use of four household appliances (refrigerator, deepfreezer, washing machine, dishwasher).

- attitudes toward energy scarcity: seven items, five-point scale measuring subjects' general attitudes toward energy scarcity.

- attitudes toward household energy conservation: the Fishbein model (Fishbein & Ajzen, 1975; Ajzen & Fishbein, 1980) was used to predict subjects' behavioral intention to conserve energy in the 1980-81 heating season. Forty-six, five-point items were included to measure subjects' beliefs about positive and negative consequences of energy conservation, their subjective evaluations of these outcomes, their normative beliefs about energy conservation, as well as their motivation to comply with these normative beliefs. The model was tested in a pilot study among a sample of 112 housewifes (cf. Ester & De Boer, 1980). Salient beliefs and consequences were derived from group discussions about energy conservation with housewifes (cf. van Amstel, Ester, van Schijndel & Schreurs, 1980). The model will be discussed in more detail in Chapter 6.

- attitudes toward mandatory or voluntary conservation: five items were selected to measure subjects' attitude toward mandatory and voluntary energy conservation options. Again, a five-point rating scale was used.

- attitudes toward the effectiveness of energy conservation information and biweekly and monthly feedback: subjects were asked whether or not they favor specific information about possibilities for residential energy conservation, as well as regular (biweekly and monthly) feedback on their household energy use.

- attitudes toward who is to blame for the energy problem and who should make conservation efforts: subjects were asked to indicate who they see as the main cause of a threatened energy scarcity situation and who should make significant conservation efforts.

5.8.2 The post-experimental questionnaire

Since subjects were explicitly informed that this interview would be considerably shorter than the first interview - which probably was a major factor in their decision to permit a second interview - only a limited number of questions could be addressed in the post-experimental questionnaire. Therefore, the questionnaire had to focus on a few selective items.
The fully structured 11-page questionnaire consisted of two parts: the first part contained questions with respect to possible attitude and behavior changes compared to subjects' pre-experimental energy attitudes and behaviors, whereas the second part addressed a number of items related to the experimental conditions subjects were assigned to.
The first part of the questionnaire dealt with the following topics:

- changes in demographic variables and household situation: number of persons, regular absence in experimental period, household composition.

- changes in energy behavior: day- and night-time thermostat setting, in experimental period, household commitments on thermostat setting, reported changes in 31 energy-relevant household behaviors, energy conservation information-seeking behavior, self-monitoring of household energy use, use of heating system.

- energy knowledge: a selection of questions from the pre-experimental interview as well as from the conservation leaflet which subjects in the experimental groups received.

- attitudes toward energy scarcity: same scale as used in the pre-experimental questionnaire.

- attitudes toward household energy conservation: due to time and space constraints as mentioned above it was not possible to use the extensive Fishbein model again. Instead it was decided to select a few items with respect to perceived success of energy conservation efforts, attitude toward conservation based on past experiences, and behavioral intention to continue with energy conservation.

- perceived comfort: four items (five-point rating scale) were selected to measure subjects' attitudes toward perceived comfort with respect to household energy consumption.

- specific items: consulting other people in neighborhood about energy conservation, household discussions about conservation, communication with neighbors about this study.

The second part of the study dealt with specific topics related to the experimental group subjects belonged to:

- Information Group: questions were addressed with regard to having read and kept the conservation booklet, frequency of consulting the information, perceived helpfulness of the booklet, attitude toward large-scale distribution of the conservation booklet. (4)

- Biweekly and Monthly Feedback Group: topics were raised with respect to having read the feedback letters, comparisons made between feedback letters, capability of explaining consumption changes, perceived helpfulness of feedback, evaluation of feedback frequency, household discussions raised by feedback letters, attitude toward large-scale implementation of feedback by utility companies and toward continuation of feedback.

- Self-monitoring Group: questions were addressed regarding recording frequency, understandability of self-monitoring forms, perceived helpfulness of self-monitoring, capability of understanding shifts in household energy consumption, family discussions raised by self-monitoring outcomes, attitudes toward continuation of self-monitoring activities and toward large scale distribution of self-monitoring recording forms.

5.9 Graphic representation of research variables

In this final section of this chapter an attempt will be made to present a graphic description of the main variables included in this study. (5)

FIGURE 5.1
RESEARCH VARIABLES

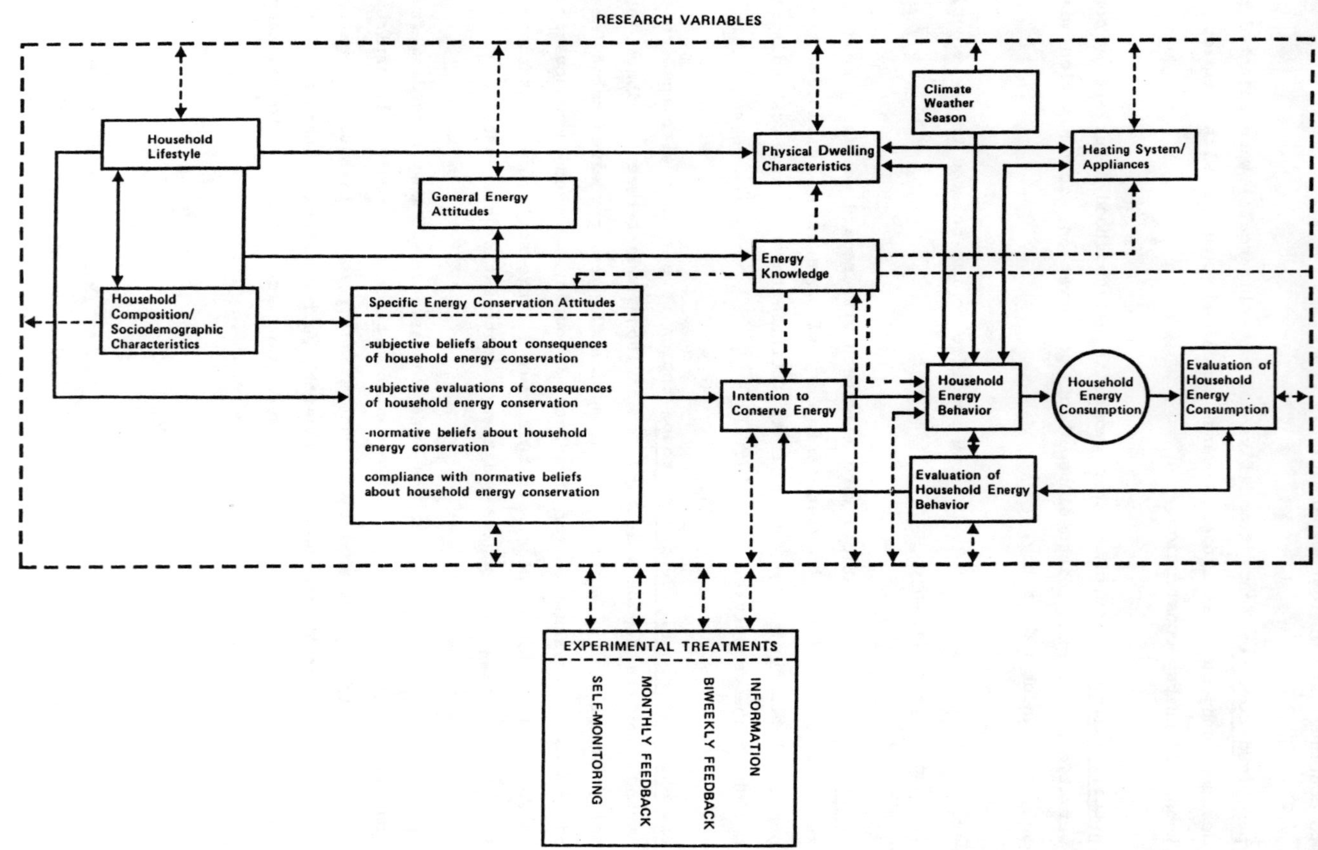

Notes

1. For further details see de Boer and Ester (1982).
2. ibid.
3. Other dwelling characteristics (e.g. sun orientation, location of dwelling, technical dwelling features) were already known since construction drawings were available of these research locations.
4. Since all experimental groups received the conservation booklet, these questions have been asked to all subjects assigned to these groups.
5. This model is an adjusted version of a behavioral model of residential energy use developed by van Raaij and Verhallen, (1983).

PART III RESULTS

6. ENERGY ATTITUDES, SOCIAL NORMS AND INTENTION TO CONSERVE ENERGY: RESULTS OF THE PRE-EXPERIMENTAL SURVEY

6.1 Introduction

This chapter will present the empirical findings from the pre-experimental survey with respect to the relationship between (general and specific) energy attitudes, social norms, and subjects' intention to conserve energy. Also, the role of energy knowledge will be investigated. As such, these findings pertain to the hypotheses (H_1, H_6, H_7, H_8) formulated in section 4.3.1 and 4.3.2 concerning this relationship and serve - among other things - to explore subjects' pre-experimental attitudinal structures regarding household energy conservation. The remaining hypotheses (H_2, H_3, H_4, H_5, H_9) will be tested in subsequent chapters of this study.
First of all, an attempt will be made to explain the intention to conserve energy through attitudinal measures by applying the Fishbein model (specific energy attitudes) and attitudes toward energy scarcity (general energy attitudes) (Section 6.2). Next, subjects' knowledge of residential energy matters will be investigated (Section 6.3) as well as their attitudes toward topics related to the experimental conditions to which subjects will be assigned in this study (Section 6.4). Then, the relationship between the above-mentioned variables and some sociodemographic characteristics will be examined (Section 6.5) and additional information will be provided about subjects' attitudes toward more general energy issues (Section 6.6). Finally, the main conclusions of this chapter will be summarized (Section 6.7).

6.2 Explaining the intention to conserve energy by specific and general energy attitudes

6.2.1 The Fishbein model

In Section 2.3.2 it was said that corresponding levels of specificity between attitudinal and behavioral measures are a necessary condition for high attitude -behavior correlations. Based on this methodological principle a number of hypotheses was offered in Section 4.3.1 with respect to the relationship between energy attitudes, the intention to conserve energy, and energy behaviors.
In this study, the Fishbein attitude-behavior model (Ajzen & Fishbein, 1980; Fishbein & Ajzen, 1975) has been applied to explore this relationship, since it explicitly deals with this necessity of corresponding levels of specifity.

Moreover, this model has been widely used and validated in social psychological research, for example with respect to family planning (Davidson & Jaccard, 1975), voting behavior (Fishbein & Coombs, 1974), blood-donating behavior (Pomazal & Jaccard, 1976), public reaction to energy proposals (Bowman & Fishbein, 1978), environmentally-conscious behavior (van der Meer, 1981), and alcohol use (Schlegel, Crawford & Sanborn, 1977), though a number of critical issues about the model has been raised too (Anderson & Shanteau, 1977; Mitchell & Biglan, 1971; Schwartz & Tessler, 1972; Songer-Nocks, 1976). Recently the model has also been applied to energy conservation behavior (Bosma & Kok, 1982; Kok, 1981; Macey & Brown, 1983; Ritsema, Midden & van der Heijden, 1982; Stutzman & Green, 1982).

According to Fishbein's "model of reasoned action", a person's behavior is assumed to be a function of the intention to perform that behavior, which, in turn, is a function of two factors: (1) attitude toward that behavior (A_{act}) and (2) subjective norms with respect to performing that behavior (SN). In turn, A_{act} is a function of beliefs about the consequences of performing the behavior and one's evaluation of those consequences, whereas SN is a function of beliefs about normative expectations with respect to the behavior and one's motivation to comply with those expectations.

In formula:

$$B \sim BI = (A_{act})w_1 + (SN)w_2 \qquad (1)$$

Furthermore:

$$A_{act} = \left[\sum_{i=1}^{n} B_i E_i\right] \qquad (2)$$

and:

$$SN = \left[\sum_{i=1}^{n} NB_i MC_i\right] \qquad (3)$$

Consequently, equation (1) can also be written as:

$$B \sim BI = \left[\sum_{i=1}^{n} B_i E_i\right]w_1 + \left[\sum_{i=1}^{n} NB_i MC_i\right]w_2 \qquad (4)$$

Where: B = overt behavior

BI = behavioral intention

B_i = beliefs about consequences

E_i = evaluation of consequences

NB_i = normative beliefs

MC_i = motivation to comply with normative beliefs

W_1 and W_2 = standardized (beta) weights to be determined by regression analysis

Figure 6.1 contains a graphic representation of the relations among those factors.

Figure 6.1: Relations among Components of the Fishbein model

beliefs about consequences (B_i)
evaluations of consequences (E_i)
attitude toward the behavior (A_{act})
relative importance of attitudinal and normative factors (W_1 and W_2)
behavioral intention (BI)
behavior (B)
normative beliefs (NB_i)
motivations to comply (MC_i)
subjective norm (SN)

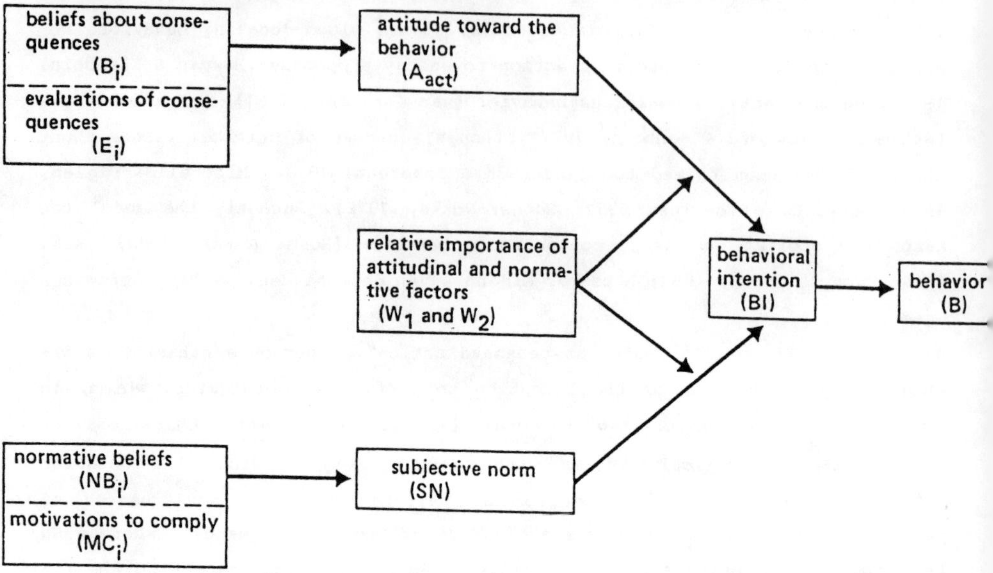

Note. Adopted from Ajzen and Fishbein (1980)

6.2.2 Raw data on the different components of the Fishbein model

6.2.2.1 Measurement format

Two pilot studies (van Amstel, Ester, van Schijndel & Schreurs, 1980; Ester & de Boer, 1980) and a previous Dutch survey study by Kok (1981) on energy attitudes and behaviors served to assess salient personal beliefs and normative expectations of the target group with regard to household energy conservation (1). Twelve personal beliefs were selected, and subjects were asked to evaluate each of the consequences and to indicate their subjective probabilities that performing the behavior (conservation by subjects on their natural gas and electricity consumption in their household) would lead to each of the consequences. Next, subjects were asked to rate the likelihood of 11 normative expectations with repect to energy conservation in their household as well as their motivation to comply with those expectations. The measurement format used was as follows. Personal beliefs (B_i) about consequences of energy conservation were stated as "If I conserve on the consumption of natural gas and electricity in my home, this would (very unlikely ... very likely) lead to ..." (selected consequence). Evaluations (E_i) of those consequences were measured through statements with the following format: "If by conserving on the consumption of natural gas and electricity in my home this would lead to ... (selected consequence), I think this is very good ... very bad" (evaluation score). Normative beliefs (NB_i) were formulated as "... (selected normative referent) thinks that I should conserve on the consumption of natural gas and electricity in my home" (very unlikely ... very likely). Motivations to comply (MC_i) with these normative expectations were measured through statements like "with respect to conservation on the consumption of natural gas and electricity I care (very much ... very little) about the opinion of ... (selected normative referent). Two direct measurements were used of the attitude toward energy conservation (A_{act}) and the subjective norm (SN) which will be explained in Section 6.2.2.4.

6.2.2.2 Beliefs about and evaluations of consequences of household energy conservation (B_i and E_i)

Table 6.1 contains subjects' mean scores on the items concerning personal beliefs and evaluations regarding consequences of household energy conservation.

Table 6.1: Mean Probabilities of Beliefs about Consequences of Household Energy Conservation and Evaluations of those Consequences

Possible Consequences of Energy Conservation in Subjects' Household	Mean Beliefs[a] (B_i)	Mean Evaluations[b] (E_i)
Less comfort for me and my family	2.6	-.04
Less environmental pollution in this country	3.5	1.5
People will think I am poor	2.1	-.08
Makes me feel more responsible for others	3.9	1.4
Lower utility bill	4.7	1.7
No increase of nuclear power plants in this country	3.3	1.5
Slow down depletion of energy supplies	3.8	1.6
People will think I am strange	2.0	-.07
Bad for national economy	1.9	-.07
Having to think a lot	2.8	.01

(Table 6.1 continued)

Possible Consequences of Energy Conservation in Subjects' Household	Mean Beliefs[a] (B_i)	Mean Evaluations[b] (E_i)
More time to develop solar and wind energy	3.0	1.3
Less cosiness at home	2.2	-1.1

Note.
[a] scales range from 1 (very unlikely) to 5 (very likely)
[b] scales range from +2 (very good) to -2 (very bad)
N = ± 468

In general, it appears from Table 6.1 that subjects have rather positive attitudes toward energy conservation, a finding which has consistently been observed in other energy survey studies (See section 2.3.1). Collective positive consequences (less environmental pollution, no increase of nuclear power plants, slow down depletion of energy supplies, more time to develop solar and wind energy) are believed to be both likely and favorable consequences of energy conservation. Personal benefits from energy conservation are also believed to be likely, whereas personal negative consequences are felt to be quite unlikely. This pattern is similar to findings from two other recent Dutch energy survey studies (Kok, 1981; Ritsema, Midden & van der Heijden, 1982).

6.2.2.3 Normative beliefs about household energy conservation and motivations to comply (NB_i and MC_i)

Table 6.2 (see page 117) presents subjects' mean probabilities of normative expectations with respect to household energy conservation and their motivation to comply with those expectations.
It is shown that, with the obvious exception of perceived normative expectations attributed to spouses/partners, the conservation norms in subjects' immediate social environment do not appear to be very salient. Both normative expectations and motivations to comply seem to be rather insignificant. At the same time, it appears that more distant normative sources (scientists, utility company, action groups, government) are assigned stronger normative conservation expectations coupled with a stronger motivation to comply with these expectations. The above-mentioned studies by Kok (1981) and Ritsema, Midden and van der Heijden (1982) also indicate the higher saliency of institutional so-

Table 6.2: Mean Probabilities of Normative Beliefs about Household Energy Conservation and Motivations to Comply with those Beliefs

Possible Normative Expectations by Several Sources to Conserve Energy in Subjects' Household	Mean Beliefs[a] (NB_i)	Mean Compliance[b] (MC_i)	N
Spouse/Partner	3.3	3.5	371
Child/Children	2.2	2.6	300
Parent/Parents	2.3	2.2	263
Friends	2.3	2.2	326
Neighbors	2.1	1.8	316
Relatives	2.1	2.0	312
Colleagues	2.0	2.0	237
Scientists	3.9	2.8	392
Utility Company	3.4	3.0	420
Action Groups	3.9	2.7	383
Government	4.3	2.9	422

Note.
[a] scales range from 5 (very likely) to 1 (very unlikely)
[b] scales range from 5 (care a lot) to 1 (don't care at all)

cial norms compared to more personal norms with regard to household energy conservation. A factor analysis on the normative sources in Table 6.2 clearly revealed this distinction between institutional and personal conservation norms as indicated in Table 6.3

Table 6.3: Factor Analysis* on NB x MC Products

NB x MC Products	Factor 1	Factor 2	N
Spouse/Partner	.44	.34	371
Child/Children	.33	.17	300
Parent/Parents	.72	.16	263
Friends	.86	.26	326
Neighbors	.79	.22	316

(Table 6.3 continued)

NB x MC Products

	Factor 1	Factor 2	N
Relatives	.83	.23	312
Colleagues	.63	.25	237
Scientists	.27	.75	392
Utility Company	.27	.70	420
Action Groups	.25	.75	383
Government	.17	.82	422
Eigenvalue	4.9	1.3	
Explained Variance	80%	20%	

Note

* Varimax rotation

6.2.2.4 Direct measurement of the attitude toward household energy conservation (A_{act}) and social norms (SN)

In correspondence with the standard Fishbein procedure two direct measurements of A_{act} and SN were used. Four semantic differentials (five-point evaluative rating scales including good/bad, advisible/nonadvisible, sensible/senseless, desirable/nondesirable) were used to assess subjects' general evaluation or overall feeling of favorableness or unfavorableness toward household energy conservation. Evaluation scores are summarized in Table 6.4.

Table 6.4: General Attitude Toward Household Energy Conservation (A_{act})

I think that conservation on my natural gas and electricity consumption is:

good	85%	5%	8%	1%	1%	bad
nonadvisable	4%	2%	6%	6%	83%	advisable
sensible	88%	3%	3%	3%	3%	senseless
nondesirable	3%	1%	7%	6%	84%	desirable

It is obvious from Table 6.4 that subjects' overall attitude toward household energy conservation is very positive, a finding which has consistently been observed in studies which used similar A_{act} measurements (Kok, 1981; Ritsema, Midden & van der Heijden, 1982).(3) It will be clear that this considerable consensus among this sample with respect to the favorableness of energy conservation offers little additional opportunity for further differentiation and analysis.

Next, subjects' subjective norm (SN) regarding household energy conservation was assessed by measuring their probability estimates of the correctness of the following statement: "most people who are important to me think I should conserve on natural gas and electricity consumption in my home". Table 6.5 contains the frequency distribution of subjects' probability scores.

Table 6.5: Subjective Norm Toward Household Energy Conservation (SN)

Most people who are important to me think I should conserve on natural gas and electricity consumption in my home:						
very likely	22%	11%	25%	7%	35%	very unlikely

As Table 6.5 shows, there is much more differentiation in subjects' probability estimates of perceived normative expectations concerning their energy consumption, than is observed in their overall attitude toward energy conservation. A similar finding is reported in a recent Dutch survey study on insulation behavior (Bosma & Kok, 1982).

6.2.2.5 <u>Behavioral intention to conserve energy in the household (BI)</u>

A crucial element in Fishbein's attitude-behavior model is the intention to perform a particular behavior (BI). In this study it was not feasible to measure behavioral intentions for a large number of energy conservation behaviors, nor to measure separate B x E's and NB x MC's for each behavior. Therefore, it had to be decided to measure subjects' general intention to conserve energy. The intention statement was formulated as: "I intend to conserve on natural gas and electricity consumption in my home this winter" (five-point scale ranging from very likely to very unlikely). Probability estimates are included in Table 6.6.

Table 6.6: Behavioral Intention to Conserve Energy in the Household (BI)

I intend to conserve on natural gas and electricity consumption in my home this winter:						
very likely	53%	15%	15%	5%	12%	very unlikely

Table 6.6. demonstrates that the majority of subjects believes it to be highly likely that they will perform energy conservation behavior in the winter season following the interview period (September/October 1980), and only a small minority feels this is unlikely.
It may therefore be concluded that the overall behavioral intention to conserve energy is quite positive in this sample.

6.2.3 Explaining A_{act} and SN by B x E and NB x MC

A first step in the analysis of the different components of the Fishbein model consists of the prediction of the attitude toward the behavior in question (A_{act}) from salient beliefs times outcome evaluations (B x E) and the prediction of subjective norms (SN) from normative beliefs times motivations to comply (NB x MC) (See Figure 6.1).

Table 6.7 contains the results of a multiple regression analysis of A_{act} on B x E.

Table 6.7: Multiple Regression of A_{act} on B x E

B x E items	Multiple R	R^2	Simple R
Less comfort for me and my family	.16	.02	.16
No increase of nuclear power plants	.24	.06	.18
People will think I am strange	.25	.06	.07
Bad for national economy	.26	.07	.01
Having to think a lot	.26	.07	.12
Lower utility bill	.46	.22	.43
Less cosiness at home	.46	.22	.11
People will think I am poor	.47	.22	-.02

(Table 6.7 continued)

B x E items	Multiple R	R^2	Simple R
Makes me feel more responsible to others	.47	.22	.25
Less environmental pollution	.48	.23	.27
More time to develop solar and wind energy	.48	.23	.26
Slow down depletion of energy supplies	.48	.23	.27

Table 6.7 shows that 23% of the variance in A_{act} can be explained through subjects' beliefs about consequences of household energy conservation and their evaluations of those consequences. Although it is evident that this result has to be characterized as only moderately successful, other attempts yielded similar findings. In the Ritsema et al. (1982) study, 27% of the variance in A_{act} could be explained through B x E measures, and 32% in the Kok (1981) study.(4) As indicated in Table 6.7 beliefs and evaluations concerning energy conservation consequences related to personal comfort, and financial savings seem to be important determinants of A_{act}. (5)

Next, we will turn to the relationship between SN and NB x MC. Table 6.8 contains the results of a multiple regression analysis.

Table 6.8: Multiple Regression of SN on NB x MC

NB x MC items	Multiple R	R^2	Simple R
Spouse/Partner	.35	.12	.35
Action Groups	.39	.15	.29
Child/Children	.42	.18	.28
Colleagues	.43	.18	.26
Parent/Parents	.43	.19	.27
Utility Company	.44	.19	.28
Scientists	.44	.19	.26
Neighbors	.44	.19	.30
Government	.44	.20	.27

(Table 6.8 continued)

NB x MC items	Multiple R	R^2	Simple R
Relatives	.45	.21	.35
Friends	.46	.21	.36

According to Table 6.8, 21% of the variance in SN can be explained by subjects' normative beliefs about household energy conservation and their motivations to comply with those normative expectations. Again it has to be emphasized, that a large proportion of this variance in SN remains unexplained. Similar attempts made by Kok (1981) and Ritsema et al. (1982) yielded percentages of explained variance of 19% and 32%, respectively.
As shown in Table 6.8 the most important normative influences are attributed to spouses/partners, action groups, and children.

6.2.4 The General Concern with Energy Scarcity Scale (GCES)

In the preceding sections the way in which specific energy conservation attitudes have been measured in this study through application of the Fishbein model was described. Some preliminary data were presented on the different components of this model. As indicated in a number of the research hypotheses, these specific energy conservation attitudes will be compared with more general energy attitudes with respect to a number of dependent variables. In this section how these general attitudes were assessed will be described.

A seven-items, five-point Likert scale was used measuring subjects' general attitudes toward energy scarcity. After factor analyzing (Varimax rotation) those items, two statements had to be excluded from this scale. These five items explain 78% of the shared variance with an eigenvalue of 2.53 and an α of .81; corrected item-total correlations vary from .48 to .66. This scale will be labeled "general concern with energy scarcity" (GCES), and will be tested against the Fishbein model.
Table 6.9 contains mean item scores and standard deviations.

Table 6.9: Mean Item Scores General Concern with Energy Scarcity Scale (GCES)

Items	M	SD
I think that all the talking and writing about oil shortages makes people more worried than justified	2.8	1.5
If you ask me, things won't take that turn with energy scarcity	3.1	1.5
Because we have sufficient natural gas resources we don't have to bother about energy scarcity	3.2	1.5
I am convinced that energy shortages will turn out less serious than expected	2.9	1.4
I don't understand why some people should always talk about energy scarcity	3.0	1.4

Note.
Scale ranges from 1 (completely agree) to 5 (completely disagree)

Table 6.9 indicates that mean item scores tend to take a medium position on the score range, demonstrating that on an average subjects are somewhat undecided about the seriousness of energy scarcity, and do not tend to favor extreme beliefs. An inspection of the standard deviations, however, reveals that there is considerable disagreement in this respect. Although there appears to be a clear segment which seems to agree with the seriousness of energy scarcity, there also is a segment which is much less concerned and feels the problem is being exaggerated. Evidence from U.S. public opinion energy surveys tends to confirm this picture (See Section 2.3.1; Olsen, 1981). In Section 6.5 an attempt will be made to further differentiate this picture by looking at some demographic correlates.

6.2.5 Relative explanatory power of specific and general energy attitudes

In Hypothesis one (H_1) it was assumed that specific energy conservation attitudes (as measured through the Fishbein model) are better predictors of the intention to conserve energy than general energy attitudes (as measured through the General Concern with Energy Scarcity Scale). Table 6.10 presents the results of a multiple regression analysis of both specific and general energy attitudes on the intention to conserve energy.

Table 6.10: Multiple Regression of BI on $\Sigma B \times E$, $\Sigma NB \times MC$, A_{act}, SN and GCES

	Multiple R	R^2	Simple R
$\Sigma B \times E$	.43	.18	.43
$\Sigma NB \times MC$	.46	.22	.29
A_{act}	.50	.25	.32
SN	.54	.30	.36
GCES	.55	.30	.14

Table 6.10 shows that the single components of the Fishbein model all correlate stronger with the intention to conserve energy than subjects' general concern with energy scarcity (GCES). It can also be concluded that including GCES in the regression analysis does not improve the prediction of BI. Thus, H_1 cannot be rejected.

It has to be added, however, that the attempt to explain subjects' intention to conserve energy by their beliefs about consequences of household energy conservation, their evaluations of those consequences, their normative beliefs with respect to conservation, their motivations to comply with those beliefs, as well as direct measurements of their attitude toward household energy conservation and subjective norms, has only been moderately successful. Together these factors explain 30% of the variance in BI. However, similar attempts have shown comparable results. By also applying the Fishbein model to measure specific energy conservation attitudes, both Kok (1981) and Ritsema, Midden, and van der Heijden (1982) explained 24% of the variance in intentions to conserve energy, whereas Stutzman and Green (1982) could explain 21%.

It could be concluded that behavioral intentions to conserve energy are complexly determined and therefore difficult to explain. Several factors might be responsible for the percentage of unexplained variance in BI in this study. First, it could be that we did not detect salient personal and normative beliefs concerning household energy conservation, though a pilot study using group interviews served to identify those salient beliefs. Second, inclusion of situational variables could improve the prediction of BI. Third, from a methodological point of view it would be better to specify the different components of the Fishbein model for separate behavioral intentions. However, as indicated in Section 6.2.2.5, this was not a feasible option in this study for obvious practical reasons.

Apart from these considerations, the following line of argument might also be clarifying in understanding determinants of behavioral intentions to conserve energy.
Recent developments in attitude-behavior theory have convincingly argued that the process of attitude formation is central to the issue of attitude-behavior relationships (cf. Fazio & Zanna, 1981). More specifically, it has been argued that the formation of an attitude is related to past behavioral experience with the attitude object. Research evidence exists which supports the hypothesis that attitudes based on direct behavioral experience with an attitude object are more predictive of behavioral intentions and later behavior than attitudes based on indirect, nonbehavioral experience (Fazio, Chen, McDonel & Sherman, 1981; Fazio & Zanna, 1978a, 1978b; Regan & Fazio, 1977). The more familiar a person is with a particular behavior through direct past experience, the more likely this person is to engage in the behavior again.
Since subjects' past behavioral experience with household energy conservation was only marginally taken into account in this study, this may also explain the relatively weak correlations between energy conservation attitudes and the intention to conserve energy. In a recent residential energy conservation study by Macey and Brown (1983) the above-mentioned hypothesis was partially confirmed. It was observed that for a number of behavioral intentions, past experience with energy conservation acts makes a significant contribution to the explanation of behavioral intentions to conserve energy above and beyond the contribution of attitudes and subjective norms (both measured through the Fishbein model).

6.2.6 Attitudinal vs. normative factors

According to Hypothesis six (H_6), attitudinal factors are stronger determinants of consumers' intention to conserve energy than normative factors. There are two ways of testing this hypothesis. First, the relative explained variance in BI by the direct measurements of A_{act} and SN could be compared. Second, the same analysis could be done for $\Sigma B \times E$ and $\Sigma NB \times MC$. Since the frequency distribution of A_{act} (See Table 6.4) showed hardly any differentiation, the second alternative offers more opportunity.
As Table 6.10 demonstrated, $\Sigma B \times E$ correlates .43 ($\underline{p}$ = .001) and $\Sigma NB \times MC$ .29 ($\underline{p}$ = .001) with BI. The regression analysis indicates that 17% of the variance in BI is explained by $\Sigma B \times E$, whereas $\Sigma NB \times MC$ only yields an additional percentage of explained variance of 4%. Beta weights for $\Sigma B \times E$ and $\Sigma NB \times MC$ are .27 and .09, respectively. Thus, although our attempt to explain differences in

BI by differences in these two attitudinal and normative factors has only been moderately successful, there is no need to reject H_6. It seems, that normative expectations and motivations to comply with those expectations with respect to household energy conservation are not very salient. This might be related to the fact that residential energy consumption is, at least to a large degree, nonvisible and anonymous behavior and therefore hardly subject to external normative control (Midden & Ritsema, 1982). A number of other behavioral energy studies have reached similar conclusions (Kok, 1981; Macey & Brown, 1983; Ritsema, Midden & van der Heijden, 1982).(6)

6.2.7 Beliefs and evaluations of personal and social consequences of household energy conservation

Hypothesis seven (H_7) predicted that consumers' intention to conserve energy will be more influenced by their beliefs and evaluations of personal consequences of energy conservation than by their beliefs and evaluations of possible social consequences. In order to test this hypothesis, subjects' internal attitude structure with respect to energy conservation consequences will be explored first to see if both dimensions are in fact represented. Table 6.11 contains the results of a factor analysis on the B x E products.

Table 6.11: Factor Analysis* on B x E Products

B x E products	Factor 1	Factor 2	Factor 3
Less comfort for me and my family	.15	.30	.03
Less environmental pollution in this country	.61	.05	-.03
People will think I am poor	-.08	.68	.04
Makes me feel more responsible for others	.59	.17	.01
Lower utility bill	.58	-.03	-.05
No increase of nuclear power plants in this country	.58	-.03	.26
Slow down depletion of energy supplies	.72	.12	-.01
People will think I am strange	-.08	.47	-.05
Bad for national economy	.01	.03	.69
Having to think a lot	.26	.30	.20

(Table 6.11 continued)

B x E products

	Factor 1	Factor 2	Factor 3
More time to develop solar and wind energy	.60	.07	.12
Less cosiness at home	.29	.39	.04
Eigenvalue	2.6	1.0	.57
Explained variance	63%	24%	14%

Note.

* Varimax rotation

Table 6.11 confirms the general assumption that two basic dimensions underlie subjects' beliefs structure with respect to consequences of household energy conservation: beliefs about social consequences (factor 1) and beliefs about personal consequences (factor 2). There are two exceptions. First, a lower utility bill as a result of energy conservation loads on the first factor and not, as one would expect, on the second factor. Second, energy conservation impacts on the national economy do not load on the first factor, but on a third one. The general picture, however, is clear and has been observed in several other Dutch energy survey studies (Bosma & Kok, 1982; Kok, 1981; Ritsema, Midden & van der Heijden, 1982).

Based on the results from this factor analysis, the following scales have been constructed. The first scale was labeled Collective advantages of energy conservation (α = .76) and refers to positive social consequences of conservation (less environmental pollution, social responsibility, no increase of number of nuclear power plants, slow down depletion of energy supplies, more time to develop solar and wind energy). The second scale was called Personal disadvantages of energy conservation (α = .48) which refers to negative individual consequences of conservation (less comfort, negative social stigmas, having to think a lot, less cosiness at home). Because of our special interest in subjects' beliefs and evaluations of personal financial consequences, the lower utility bill item will be looked at separately in subsequent analyses. (7)

If H_7 is true, one would expect beliefs about personal consequences of household energy conservation to be stronger correlated with the intention to conserve energy than beliefs about social consequences. Table 6.12 contains the correlation matrix of these variables.

Table 6.12: Correlation Matrix of Beliefs about Personal and Social Consequences of Energy Conservation and the Intention to Conserve Energy

	Collective Advantages	Personal Disadvantages	Personal Financial Consequences	BI
Collective advantages				
Personal disadvantages	-.29**			
Personal financial consequences	.50**	-.11*		
BI	.34**	-.34**	-.28**	

* p ≤ .01

** p ≤ .001

Table 6.12 does not support H_7. Beliefs about personal impacts of household energy conservation do not yield higher correlations with subjects' intention to conserve energy, compared to beliefs about collective or social impacts. One has to realize, however, that strictly speaking the measurement of beliefs and evaluations did not include a subjective hierarchy in salience of those beliefs and evaluations, which would permit a stronger test of the hypothesis. Nevertheless, it has to be concluded from these data that the intention to conserve energy correlates about equally strong with personal and social motives, as well as that those motives tend to be interrelated.

6.3 Energy knowledge

A number of social energy studies have observed the phenomenon of consumer "energy illiteracy" (See section 2.3.3), indicating that in general consumers have low levels of energy knowledge.

In Hypothesis eight (H_8) it was hypothesized that there is a significant relationship between consumers' knowledge of residential energy matters and their specific and general energy attitudes and their intention to conserve energy. In this section the degree of empirical support for this hypothesis will be determined.

In order to test this hypothesis, a scale was developed to measure subjects' knowledge of residential energy matters. This scale consists of six questions: knowledge of the price of one cubic meter natural gas and one KWH electricity, knowledge of one's last year's consumption of natural gas and electricity, and two questions about insulation subsidies. (α of this scale is .58).

Table 6.13 shows the number of knowledge questions subjects correctly answered.

Table 6.13: Percentage of Subjects Correctly Answering Energy Knowledge Questions

Correctly answered questions	%
zero questions	24%
one question	49%
two questions	17%
three questions	6%
four questions	3%
five questions	1%
six questions	1%

Table 6.13 supports the notion of "energy illiteracy". About one-fourth of the subjects did not know the correct answer to any of the knowledge items, 50% answered only one item correctly, 17% two items, and only 11% gave correct answers to three or more knowledge items. Thus, it appears that in correspondence with the findings from similar studies reported in Section 2.3.3, the sample is not very informed about residential energy matters.

Table 6.14 reports the findings with respect to H_8. For a correct understanding of the data, it has to be noticed that in this table - as well as in subsequent analyses - results will be reported for the three factors (collective advantages, personal disadvantages, personal financial consequences) observed in $\Sigma B \times E$, rather than $\Sigma B \times E$ itself, and the two factors (institutional norms, personal norms) observed in $\Sigma NB \times MC$, rather than $\Sigma NB \times MC$ itself.

Table 6.14: Correlations Between Energy Knowledge and Specific and General Energy Attitudes and the Intention to Conserve Energy

	Energy knowledge
General concern with energy scarcity (GCES)	.15**
A_{act}	.17**
SN	.02
Collective advantages	.18**

(Table 6.14 continued)

	Energy knowledge
Personal disadvantages	-.08
Personal financial consequences	.15**
Personal norms	.05
Institutional norms	.05
BI	.14*

* p = $\leq$.01
** p = $\leq$.001

Although the correlations reported in Table 6.14 obviously are of a rather modest nature, H_8 has to be rejected nevertheless. It appears that contrary to this hypothesis, knowledge of residential energy matters is somewhat related to general concern with energy scarcity, A_{act}, beliefs about personal and collective consequences of energy conservation, as well as to the intention to conserve energy. Energy knowledge, however, is not significantly related to SN, nor to personal or institutional norms with respect to energy conservation. Thus, though H_8 is not supported one has to bear in mind that energy knowledge is only weakly related to the different components of the Fishbein model (and to general energy attitudes). This conclusion was also reached in a recent study by Stutzman and Green (1982).

6.4 Attitudes related to the experimental conditions

In section 4.3.4 a number of hypotheses was formulated with respect to the relationship between the effectiveness of behavioral interventions tested in this study, and subjects' pre-experimental need for energy conservation information and regular feedback. This section presents some raw data on those needs. First, subjects were asked to indicate whether they had sufficient or insufficient energy conservation information, and next, whether concrete and practical conservation information would encourage them to conserve energy in their households.

Table 6.15: Subjects' Pre-Experimental Need for Energy Conservation Information

	agree completely	agree	don't agree/ don't disagree	disagree	disagree completely	have enough information	already conserve enough
I have insufficient energy conservation information	24%	24%	7%	18%	27%		
If I would have concrete and practical energy conservation information, I would certainly conserve energy	45%	18%	6%	6%	7%	12%	7%

As Table 6.15 shows, roughly one-half of the sample somewhat or completely agrees with the first statement of having insufficient energy conservation information, whereas the other half believes the opposite to be the case. About sixty percent feels that having concrete and practical energy conservation information would certainly stimulate them to conserve energy in their households; 12% believes they already conserve enough, however. Not surprisingly, both variables are strongly interrelated (R = .53, $\underline{p}$ = .001).
Next, subjects were asked whether providing them with biweekly or monthly feedback on their household energy consumption would help them to conserve energy.

Table 6.16: Subjects' Pre-Experimental Need for Energy Consumption Feedback

	would help very much	would help	would help little	would not help	don't know	I don't want to conserve energy	impossible to conserve more energy
Biweekly feedback	21%	25%	19%	28%	3%	1%	3%
Monthly feedback	21%	30%	15%	26%	3%	1%	4%

As Table 6.16 demonstrates, subjects tend to have somewhat mixed feelings about energy consumption feedback. The segment that feels that feedback would help somewhat or very much to conserve energy is about equally large as the segment that feels it would help little or not at all. There is hardly any difference in subject's beliefs about biweekly compared to monthly feedback, which is reflected in the strong correlation between those two beliefs ($R = .85$, $p < .001$).

No statistically significant relationships were found between subjects' needs for energy conservation information and feedback, and their behavioral intention to conserve energy or their energy knowledge. (8)

6.5 Sociodemographic correlates

This section will briefly look at the relationship between variables described in previous sections of this chapter and some sociodemographic characteristics. The analysis will be restricted to subjects' age, education, and socioeconomic status. (9) Findings reported in this section may provide some basic insight into sociodemographic correlates of energy conservation attitudes, social norms, conservation intentions, and energy knowledge, as well as of subjects' pre-experimental needs for energy conservation information and regular energy feedback. In addition, these findings may generate segmentation criteria for further analyses in subsequent chapters. Results are summarized in Table 6.17.

Table 6.17: Sociodemographic Correlates of Specific and General Energy Attitudes, Intention to Conserve Energy, Energy Knowledge, and Needs for Energy Conservation Information and Energy Consumption Feedback

	Age	Education	Socioeconomic Status
General concern with energy scarcity (GCES)	-.20***	.25***	.15***
A_{act}	-.08*	.12**	.16**
Collective advantages	-.08	.11**	.09*
Personal disadvantages	-.08*	.07	.08*
Personal financial consequences	-.12**	.17***	.14***
SN	.06	.02	.03

(Table 6.17 continued)

	Age	Education	Socioeconomic Status
Personal norms	-.17***	.12**	.16***
Institutional norms	-.14**	.09*	.05
BI	-.15***	.12**	.13**
Energy knowledge	-.16***	.17***	.24***
Insufficient energy conservation information	.03	-.21***	.13**
Perceived effectiveness of energy conservation information	-.11**	-.12**	-.05
Perceived effectiveness of biweekly feedback	-.03	-.08*	-.13**
Perceived effectiveness of monthly feedback	-.07	-.06	-.08*

* $p \leqslant .05$
** $p \leqslant .01$
*** $p \leqslant .001$

Obviously, the overall conclusion from Table 6.17 is that sociodemographic characteristics are only moderately or weakly correlated with specific and general energy attitudes, conservation intentions, energy knowledge, and perceived effectiveness of energy conservation information and feedback. This finding, however, has been observed in many behavioral energy studies (See section 2.3.4).

There is some evidence that as far as age is concerned, younger subjects compared to older subjects are somewhat more concerned with energy scarcity, less convinced with negative personal financial consequences of energy conservation, less sensitive to both personal and institutional conservation norms, more willing to conserve energy, have higher levels of energy knowledge, and perceive more effectiveness of energy conservation information.

Higher educated subjects are more concerned about energy scarcity, have more favorable attitudes toward energy conservation, are more convinced of personal positive financial consequences of household energy conservation, more sensitive to personal conservation norms, more willing to conserve energy, have higher knowledge levels, believe more that they have sufficient conservation information, and perceive somewhat more effectiveness of conservation information than lower educated subjects. Largely, the same conclusions hold for socioeconomic status.

Again, those findings and conclusions have to be interpreted with some scepticism given the overall low correlations reported in Table 6.17.

According to the Fishbein model, external variables like sociodemographic characteristics are only indirectly related to behavioral intentions and are mediated through beliefs underlying the intention (cf. Ajzen & Fishbein, 1980, p. 82-91). After partialling out those beliefs, the correlations between behavioral intention and age, education, and socioeconomic status no longer reach statistically significant levels.

6.6 Attitudes toward other energy issues

A number of questions in the pre-experimental questionnaire were related to some more general energy issues. Although those issues are only indirectly relevant for the main theme of this study, they may nevertheless provide some further information about the psychological make-up of the sample with respect to the energy question. Findings will just be briefly summarized.

When asked who is to blame in the first place for energy scarcity, 22% of the subjects blames the industry, 21% the OPEC-countries, 17% the government, 12% the oil companies, 11% the citizens, 13% does not know, and 5% feels there is no threatening energy scarcity (cf. Cunningham & Lopreato, 1977; Millstein, 1976). Subjects were also asked how much energy should be conserved by a number of actors (industry, government, citizens, respondent itself). Sixty-five percent believes that industry should conserve considerably, and 80% feels the same holds for government. Although 50% of the subjects says citizens should conserve much or very much, a notable lower percentage (35%) feels that this also applies to the subject itself.

An important policy topic is whether consumer energy conservation should be based on voluntariness or on coercion. About 70% of the subjects is in favor of voluntary energy conservation, whereas 16% believes government should force consumers to conserve energy. A number of questions built on this theme. A majority of subjects (60%) believes that consumers should decide for themselves whether or not - and how much - to conserve energy, even though 40% doubts whether people are voluntarily willing to conserve. Although about half of the subjects believes that voluntary energy conservation has proven to be succesful, seventy percent nevertheless feels that more control is needed to make sure that everyone contributes.

Finally, more than two-thirds thinks that rising energy prices will automatically lead consumers to conserve energy.

6.7 Summary

The attempt made in this study to explain subjects' intention to conserve energy by their beliefs about consequences of energy conservation and their evaluation of those consequences, as well as beliefs about normative expectations, has only been moderately successful. Together these specific energy attitudes explain 30% of the variance in the intention to conserve energy. Apparently, energy conservation intentions are complexly determined and difficult to predict. Four suggestions were offered to improve this prediction: (a) better measurement of salient energy conservation beliefs, (b) more precise inclusion of situational variables, (c) unfolding of conservation intentions within separate behavioral contexts, and (d) inclusion of subjects' past behavioral experience with energy conservation (cf. Stern, Black & Elworth, 1982, 1983).
It was found that both the overall attitude toward household energy conservation and the intention to conserve energy are quite positive in our sample.
The hypothesis (H_1) that specific energy conservation attitudes are better predictors of intentions to conserve energy than general energy attitudes was supported. In correspondence with H_6, it was observed that attitudinal factors are stronger determinants of subjects' intention to conserve energy than normative factors. No empirical support, however, could be found for the hypothesis (H_7) that the intention to conserve energy is more influenced by subjects' beliefs about and evaluations of personal consequences of energy conservation than by their beliefs and evaluations of possible social consequences. Both factors correlated about equally strong with the intention to conserve energy.
The phenomenon of "consumer energy illiteracy" was also observed in this study. Although the correlations were generally rather weak, the hypothesis (H_8) that there is no significant relationship between consumers' knowledge of residential energy matters and their specific and general energy attitudes and their intention to conserve energy nevertheless had to be rejected.
Next, some data were provided about subjects' attitudes related to the experimental conditions, particularly their pre-experimental needs for energy conservation information and regular energy consumption feedback.
Correlational evidence suggested that specific and general energy attitudes, intention to conserve energy, energy knowledge, and needs for energy conservation information and regular feedback are only moderately or weakly related to subjects' sociodemographic characteristics (age, education, socioeconomic status). Finally, some information was provided about subjects' attitudes related to some more general energy issues.

Notes

1. See also Kok, Abrahamse, Douma, Langejan, Sietsma, Slob, and de Vries (1979).

2. In a pilot study (Ester & de Boer, 1980) it was found that separate statements about natural gas and electricity use did not discriminate and turned out to be extremely time consuming and very boring to subjects.

3. See also Section 2.3.1.

4. See also Stutzman and Green (1982).

5. As mentioned in Chapter 1, there exists a large public concern with nuclear energy in the Netherlands.

6. Stutzman and Green (1982), however, concluded that both attitudinal and normative factors (as measured through A_{act} and SN) were useful in predicting energy conservation intentions.

7. Some caution is needed here given the skewed frequency distribution of this item.

8. There are two exceptions, which are of only marginal importance, BI correlates .10 (p = .02) with perceived effectiveness of monthly feedback, and energy knowledge correlates .08 (p = .03) with perceived effectiveness of biweekly feedback.

9. See Section 2.3.4.

7. EXPERIMENTAL RESULTS

7.1 Introduction

In this chapter the main experimental findings will be discussed. First of all, however, the necessary energy consumption data preparations applied in this study will be outlined (Section 7.2), and some information will be provided about baseline consumption in the five research locations (Section 7.3).
Section 7.4 will analyze the absolute and relative effectiveness of the experimental conditions in terms of promoting energy conservation. As such, those findings pertain to the hypotheses (H_9, H_{10}, H_{11}, H_{12}, H_{13}) outlined in Section 4.3.3. In Section 7.5 the relationship between experimental outcomes and specific and general energy attitudes will be explored, with special reference to a number of hypotheses (H_2, H_4, H_{14}) mentioned in Section 4.2.1 and 4.3.4.
Next the relationship between experimental results and possible energy behavior changes adopted by subjects is investigated (Section 7.6). In this section, some data are reported in order to test Hypothesis three (H_3) and five (H_5) as mentioned in Section 4.3.2. Following, it will be analyzed whether effectiveness of behavioral interventions is related to subjects' sociodemographic characteristics (Section 7.7). Finally, the main findings reported in this chapter are summarized in Section 7.8.

7.2 Consumption data preparation

This section describes the data preparations which have been applied with respect to the energy consumption data gathered in both the pre-experimental (baseline) period, the experimental period, and the post-experimental period.

7.2.1 Standardized consumption

A first step in the data preparation procedure was to carefully inspect and check the energy consumption measures in all phases of this study in order to detect possible errors or peculiarities. Since subjects' 1980 baseline consumption as calculated from the utility company files (See Section 5.4.3) did not always cover identical consumption periods, a standardized measure of natural gas and electricity was computed for each subject by using the proportional correction methods described in Section 5.3.2. Correction for small differences in recording dates has also been applied with respect to subjects' energy consumption in the experimental and post-experimental period.

7.2.2 Weighted consumption

Residential energy use is a function of many factors (e.g. outside temperature, thermic insulation quality of dwellings, heating patterns) which interact in their effects on energy consumption. This fact implies that it is hardly possible to provide exact information about who conserved energy and to what degree, apart from the heuristic problem of which consumption changes can be classified as conservation acts. From a methodological point of view, therefore, it makes more sense to focus on energy consumption trends.
In this study, consumption trends were analyzed at two levels: the individual level and the aggregated level. The individual level corresponds to trends in energy consumption of subjects assigned to both feedback conditions (See Section 5.3.2), whereas at the aggregated level energy consumption trends between experimental conditions are compared. At the aggregated level a number of correction procedures are not required, since energy consumption of the experimental groups can be compared with the control group's consumption. However, a weighting procedure is needed because the effects of energy behavior changes are likely to be mediated by dwelling characteristics and electrical appliances present in the household. Although, research locations have been selected in this study with physically identical, mainly single-family dwellings with individually metered gas-fired central heating systems, some differences nevertheless remain (e.g. corner or middle dwellings, sun and wind orientation).
The following weighting procedure has been applied to account for those differences with regard to natural gas and electricity consumption data, respectively.

As far as natural gas is concerned, differences in physical dwelling characteristics (position of dwelling, small variations in dwelling type, sun and wind orientation) were used as dummy variables in a regression analysis of natural gas consumption per research location. Also, natural gas use of the total sample was related to additional information obtained about transmission and ventilation losses of the dwellings concerned.
Next, the outcome of the analysis for the total group has been checked again per location to see if all relevant information of the dummy variables had been accounted for. Finally, the resulting regression equation was used to weight the natural gas consumption measures. By computing the natural logarithm of all consumption measures, an estimate is obtained of the fraction a particular subject is using more of less than other subjects in a comparable dwelling situation.
Weighted electricity consumption indicates which fraction one is using more or

less than other subjects with similar appliances. Using regression analysis to predict electricity consumption based on number of electrical appliances is difficult for at least two reasons. First, functionally similar appliances may use different amounts of electricity. Second, usage intensity is a more determining factor than mere possession of appliances. In order to improve the analysis a number of questions from the pre-experimental questionnaire were used which specify usage intensity, both directly (use of washing machine) and indirectly (number of children and adults).

By including number of electrical appliances, and direct and indirect measures of usage intensity in a regression analysis, 79% of the variance in subjects' 1980 electricity consumption could be explained. It has to be added, however, that the standard deviation of the residuals still amounts to 27% of mean electricity consumption. The resulting regression equation was used to weight subjects' 1980 and 1981 electricity consumption, and, finally, the natural logarithm was computed.

7.2.3 Relative 1980-81 consumption

Most analyses in this study are related to subjects' relative 1980-81 energy consumption. This measure indicates the regression of subjects' energy consumption in 1981 on their consumption in 1980, after weighting of the factors described in the preceding section and after logarithmic transformation of the consumption measures.

The outcome of this regression analysis indicates the general trend observed in subjects' energy consumption, whereas the residuals point to increasing or decreasing deviations.

7.3 Baseline consumption in research locations

By using the proportional correction methods described in Section 5.3.2 and by accounting for differences in recording dates (See Section 7.2.1), a standardized baseline consumption for each subject was computed.

Table 7.1 (see page 140) shows the outcome of this procedure for each research location.

A number of observations can be made from Table 7.1. It turns out that mean natural gas consumption is highest in Hilversum (large, noninsulated dwellings), whereas mean electricity consumption is highest in Zaandam (dwellings are provided with an electric boiler). The Amsterdam location is characterized by relatively low levels of natural gas and electricity consumption. It can also be observed that within each research location, subjects differ consider-

Table 7.1: Standardized Mean Baseline Energy Consumption in Research Locations

Location	M	SD	N
	Natural Gas (m^3)		
Heerhugowaard	2516	622	75
Zaandam - noninsulated dwellings	2227	620	45
- insulated dwellings*	2067	521	41
Purmerend - noninsulated dwellings	2903	604	10
- insulated dwellings	2543	492	65
Hilversum	3244	605	81
Amsterdam	2074	626	70
	Electricity (KWH)		
Heerhugowaard	2438	724	75
Zaandam	4936	941	86
Purmerend	2981	933	75
Hilversum	2793	786	81
Amsterdam	2235	774	70

* See section 5.2

ably in their consumption of natural gas and electricity. Standard deviations with respect to mean natural gas consumption range from 20-30%, and are even higher - with the exception of Zaandam - with respect to mean electricity consumption. This finding has been observed in many behavioral energy studies (Sonderegger, 1978; Verhallen & van Raaij, 1979, 1980; Wotaki, 1977). (2)
A subsequent analysis indicated that the variation coefficient ($\frac{sd}{x}$ 100) per location is quite stable over the different (pre-experimental, experimental, post-experimental) consumption periods, which means that while energy consumption increases proportionally over time, the relative differences between subjects remain rather stable. As outlined in the previous sections, this finding has been used at the individual level to correct for differences in recording date and to provide subjects with feedback about trends in their energy consumption, and at the aggregated level to sharpen the analysis of experimental effects.

7.4 Effectiveness of experimental conditions

This section shows the experimental results concerning the effectiveness of behavioral interventions tested in this study. As stated in the hypotheses (H_9, H_{10}, H_{11}, H_{12}), reduced energy consumption was predicted in all experimental conditions (See Section 4.3.3). In terms of relative effectiveness (H_{13}), self-monitoring of energy use was expected to be the most effective intervention in reducing subjects' energy consumption, whereas providing energy conservation information was predicted to be least effective. It was also hypothesized that biweekly feedback is more effective than monthly feedback.

First of all, it appears that subject's weighted energy consumption during baseline is strongly related (R^2 about .80) to their energy consumption in the experimental period. Given this observed continuity in subjects' energy consumption a decreasing tendency is found. The overall (non-specified) experimental effects (reduction percentages) are shown in Table 7.2.

Table 7.2: Overall Experimental Results (Control Group Comparisons)

	Natural Gas Consumption	Electricity Consumption
Reduction in experimental period	- 3%*	-1%
Reduction in post-experimental period	- 4%*	-3%

* $p \leq .05$

Table 7.2 indicates that, after correction for the factors described previously, the four experimental groups use less energy in both the experimental and post-experimental period compared to the control group. It has to be added, however, that the reduction percentages are modest and not statistically significant for electricity consumption. Table 7.3 differentiates those results according to experimental condition.

Table 7.3: Experimental Results by Experimental Condition (Control Group Comparison)

Experimental Condition	N	Experimental Period		Post-Experimental Period	
		Natural Gas Consumption	Electricity Consumption	Natural Gas Consumption	Electricity Consumption
Conservation information	74	-3%	-1%	-5%*	-2%
Biweekly feedback	77	-4%*	0%	-3%	-2%
Monthly feedback	76	-3%	-1%	-3%	-4%
Self-monitoring	73	-4%*	-1%	-4%	-3%

* $p < .05$

A number of conclusions can be drawn from Table 7.3. First, although findings are generally in the direction predicted by H_9, H_{10}, H_{11}, and H_{12}, they appear to be statistically significant in only two instances: biweekly feedback and self-monitoring. In no condition statistically significant effects for electricity consumption could be found. Perhaps one could explain this last finding by the fact that electricity consumption, more than natural gas consumption, is the sum of a large number of quite different behaviors that cannot easily be changed. Also, a large proportion of electricity consumption is used "automatically" (e.g. appliances) and is therefore little subject to behavior change.

Next, it has to be concluded that our hypothesis about the relative effectiveness of the four behavioral interventions (H_{13}) has to be rejected. There are no substantial differences in effectiveness between interventions. Subsequently, parametric and non-parametric statistical tests do not alter this conclusion.

It is obvious that these findings on the effectiveness of feedback differ markedly from those of most other energy feedback studies, in which reduction percentages of up to 15% - 20% have been reached (See Section 3.3.2.1). The difference, however, is that those studies often used daily feedback. Though we disputed the policy relevance of highly frequent feedback, it can be concluded that less frequent -and therefore less costly - feedback is not an adequate alternative given the findings reported in Table 7.3. Perhaps infrequent feedback blocks the learning and motivational functions of feedback. In the next chapter we will come back to this cognitive explanation based on data from the post-experimental questionnaire.

In Section 3.3.1.1 it was concluded that most behavioral experiments on residential energy conservation did not find significant effects of providing subjects with energy conservation information. Typically, however, most experiments were restricted to short-term effects. It is therefore interesting that the information condition in this experiment was the only condition which produced statistically significant post-experimental reductions. Thus, taking into account longer-term effects of energy conservation information may differentiate the existing viewpoints popular among many applied behavior analysts which usually question the usefulness and effectiveness of conservation information. Our findings suggest that well-designed informational strategies focusing on concrete and specific behavior changes do stimulate conservation behavior (cf. Ester & Winett, 1982; Winett & Ester, 1983).
Self-monitoring of energy use, finally, resulted in a statistically significant reduction of energy consumption in the experimental period. Although the reduction percentage is at best modest, the obvious low-cost nature of self-monitoring confirms its practical usefulness from a policy point of view.

Despite the fact that the experimental interventions tested in this study yielded relatively small reduction percentages, the amount of cubic meters natural gas "saved", could still be considerable when taken community- or nation-wide. (3) Table 7.4 shows weighted experimental and post-experimental natural gas consumption.

Table 7.4: Weighted Experimental and Post-Experimental Natural Gas Consumption by Experimental Condition

Experimental Condition	M	SD	N
	(m^3)	(m^3)	
Conservation information	2273	465	74
Biweekly feedback	2263	498	77
Monthly feedback	2280	476	76
Self-monitoring	2260	446	73
Control group	2351	533	74

The average reduction percentages of about 3%-4% reached in this study on subjects' natural gas consumption, equal about 70 to 90 m^3 on a yearly basis. At the aggregated level such savings should not be undervalued.

Finally, it was analyzed whether subjects with a relatively high baseline energy consumption level responded differently to the experimental conditions, than subjects with a relatively low baseline consumption. No significant interaction effect could be found between research locations - which differ in energy consumption baseline levels - and responsiveness to experimental conditions. (4) Thus, there is no immediate need to assume that the experimental interventions tested in this study have been more or less effective in locations with high or low baseline consumption levels.

7.5 Experimental results and specific and general energy attitudes

Two topics will be addressed in this section. First, some findings will be presented on the relationship between specific and general energy attitudes and real energy consumption (Section 7.5.2). Next, these attitudes will be related to the effectiveness of the experimental interventions (Section 7.5.2).

7.5.1 Energy consumption and specific and general energy attitudes

In the preceding chapter the relationship between subjects' specific and general energy attitudes and their intention to conserve energy was discussed. In this section we will go one step further by relating those attitudes to real energy consumption measures. According to Hypothesis two (H_2), specific attitudes of consumers toward energy conservation are stronger predictors than consumers' general attitudes toward energy scarcity - both as measured before the heating season sets in - of their real energy consumption in that season. Hypothesis four (H_4) stated that the intention to conserve energy is a moderate predictor of consumers' energy consumption in the heating season (See Section 4.3.1).

It will be obvious that both hypotheses cannot be tested for the experimental groups, given their experimental treatments. Therefore, the analysis will be limited to the control group. The two hypotheses will be tested by including the independent variables concerned into a multiple regression analysis with subjects' relative 1980-81 natural gas consumption as the dependent variable. Findings are shown in Table 7.5.

Table 7.5: Multiple Regression of Subjects' Relative 1980-81 Natural Gas Consumption on Specific and General Energy Attitudes and the Intention to Conserve Energy

	Multiple R	R^2	Simple R
Behavioral intention	.13	.02	-.13
Personal financial consequences	.15	.02	-.12
Personal disadvantages	.22	.05	-.20
Collective advantages	.30	.09	-.28
Institutional norms	.30	.09	-.06
Personal norms	.30	.09	-.09
A_{act}	.31	.10	-.11
SN	.32	.10	-.15
General concern with energy scarcity (GCES)	.32	.10	-.05

Note.
Findings pertain to control group only (N = 77)

Obviously, one has to conclude from Table 7.5 that only a small fraction of the control group's relative 1980-81 natural gas consumption can be explained by subjects' pre-experimental specific and general energy attitudes. Only 10% of the variance in consumption is explained by those attitudinal measures. Given this outcome, however, there is some limited evidence for both hypotheses. Specific energy conservation attitudes are somewhat stronger related to subjects' relative 1980-81 gas consumption than general attitudes toward energy scarcity. This seems especially the case for beliefs about personal disadvantages and collective advantages of household energy conservation. In favor of H_4 it is found that the behavioral intention to conserve energy is only moderately related to subjects' natural gas consumption. Besides the explanations offered in Section 6.2.5 (salience of conservation beliefs, situational constraints, non-specific character of measurement of behavioral intention to conserve energy, past behavioral experience with energy conservation), it could be hypothesized that the time span between measurement of the behavioral intention to conserve energy and measurement of energy consumption is related to this relatively weak correlation (Ajzen & Fishbein, 1980, p. 34-39). The longer the time span between attitudinal and behavioral measures, the less likely are strong correlations between those measures will be found. Stutzman and Green (1982) found empirical support for this hypothesis in their study of factors affecting energy consumption. These authors concluded that behavioral intention measures decrease their potency in predicting energy usage when the time between the measurement of both variables is longer than two months. (5)

7.5.2 Effectiveness of experimental conditions and attitudes toward energy conservation

According to Hypothesis fourteen (H_{14}), the more positive consumers' pre-experimental attitudes toward energy conservation, the more effective information, feedback, and self-monitoring are in reducing their energy consumption. This was referred to as the "breeding ground hypothesis" (Section 4.3.4). This section will provide some empirical data about this hypothesis.

In section 6.2.5 some recent developments in attitude-behavior theory were discussed which focus on the process of attitude formation, primarily by studying past behavioral experience with the behavioral object. In this study, subjects' past behavioral experience with energy conservation (natural gas usage) was measured by asking whether practical agreements on thermostat setting exist in their household. (6)

Table 7.6 shows the correlations between energy conservation attitudes and subjects' relative 1980-81 natural gas consumption by existence of thermostat setting agreements. Given the observed small differences between experimental conditions, results will be reported for all four experimental groups together.

Table 7.6: Correlations Between Energy Conservation Attitudes and Subjects' Relative 1980-81 Natural Gas Consumption by Thermostat Setting Agreements

	Experimental Groups	Control Group	All Groups	Thermostat Setting Agreements	No Thermostat Setting Agreements
	N = 309	N = 77	N = 386	N = 91	N = 247
Behavioral intention	-.09	-.13	-.10*	-.17	-.06
Personal financial consequences	-.03	-.12	-.05	-.04	-.04
Personal disadvantages	-.11*	-.20*	-.13*	-.27**	-.07

(Table 7.6 continued)

	Experimental Groups	Control Group	All Groups	Thermostat Setting Agreements	No Thermostat Setting Agreements
Collective advantages	-.02	-.28**	-.07	-.26**	-.01
Institutional norms	-.02	-.06	-.03	-.04	-.02
Personal norms	.04	-.10	.01	-.17	-.06
A_{act}	.02	-.11	-.02	-.10	-.01
SN	-.08	.15	-.02	-.02	-.03
General concern with energy scarcity (GCES)	.05	-.05	.03	-.04	.09

* $p < .05$

** $p < .01$

Table 7.6 provides little empirical support for the "breeding ground hypothesis" -both the general attitude toward energy scarcity (GCES) and specific attitudes toward energy conservation are not statistically significantly related to consumption trends in the experimental groups. Only beliefs about negative personal consequences of residential energy conservation turn out to be significantly related to subjects' relative 1980-81 natural gas consumption. The strength of this correlation is, however, quite weak ($R = .11$).

An alternative explanation for this finding could be that the experimental conditions have in fact influenced subjects with relatively less positive attitudes toward energy conservation by making their attitudes more positive. As such, the experiment would confound the attitude-consumption relationship. There appears to be some evidence in favor of this argument. As Table 7.6 indicates, control group subjects show stronger attitude-consumption correlations than subjects in the experimental groups. However, a subsequent analysis revealed that the only statistically significant interaction with experimental effects could be found for the attitude toward collective benefits of energy conservation. In this case, it turned out that the experimental conditions were relatively most effective for subjects with less positive attitudes (a 6% reduction in relative 1980-81 natural gas consumption). No experimental effects could be found for subjects with positive attitudes. The same finding was ob-

served in the post-experimental period, though the interaction effect did not turn out to be statistically significant.
Since from a logical point of view these observations do not neccessarily imply energy conservation attitude change by subjects, we will address this issue in the next chapter.
Interestingly, Table 7.6 shows that when thermostat setting agreements exist in the household, specific energy conservation attitudes yield better predictions of subjects' relative 1980-81 natural gas consumption, then when such agreements do not exist. This finding might be interpreted in favor of the hypothesis stating that prior experience with energy conservation is positively correlated with the strength of the attitude-behavior relationship (Macey & Brown, 1983; Regan & Fazio, 1977).

7.6 Experimental results and energy behavior change

Two topics will be addressed in this section. First, the question of which energy behavior changes subjects have adopted in both the experimental groups and the control group, relative to the pre-experimental period will be explored (Section 7.6.1). Second, some data will be reported on the relationship between those behavior changes and subjects' specific and general energy attitudes (Section 7.6.2).

7.6.1 Energy behavior change in experimental period

Many studies have shown that environmental behavior in general and energy behavior in particular is not a set of homogeneous, highly correlated behaviors, but tends to be a concept which refers to many quite diverse and often not interrelated behaviors (Ester, 1978; Ester & van der Meer, 1982; Geller, Winett & Everett, 1982; Lipsey, 1977; van der Meer, 1981; Ritsema, Midden & van der Heijden, 1982; Rozendal, Ester & van der Meer, 1983). Therefore, energy behavior appears to be a complex and multi-dimensional concept, and attempts to empirically model this behavior have only been moderately successful.
This section will include an investigation of which reported energy behavior changes subjects have adopted during the experimental interventions. The measurement procedure has been as follows. In the post-experimental interview subjects were given a set of 31 different energy-related behaviors and were asked to indicate which behaviors they had performed more or less in the past half year. Next, they were instructed to select from this sub-set those behaviors which they had intentionally adopted for energy conservation purposes. Table 7.7 shows which energy conservation behaviors have been intentionally

adopted by more than ten percent of all subjects.

Table 7.7: Intentionally Adopted Energy Conservation Behaviors* by Experimental Condition

Conservation Behaviors	Conservation Information Group	Biweekly Feedback Group	Monthly Feedback Group	Self Monitoring Group	Control Group	All Subjects
	N = 69	N = 74	N = 71	N = 69	N = 70	N = 349
	% Adoption					
- doing the laundry at 60°C	15%	26%	23%	23%	21%	22%
- closing curtains in living-room	16%	16%	23%	15%	17%	18%
- checking thermostat setting	13%	20%	18%	19%	17%	18%
- no pre-washing	10%	16%	16%	11%	13%	13%
- using less lighting	23%	34%	25%	23%	24%	26%
- closing door between hall and living-room	10%	19%	18%	12%	10%	14%
- lowering thermostat one hour before going to bed	22%	22%	28%	23%	21%	23%
- installing lower-wattage bulbs	13%	16%	14%	8%	12%	13%
- lowering temperature in living room	9%	16%	9%	12%	7%	11%
- lowering thermostat when leaving the house	12%	19%	20%	12%	23%	17%

* *Note*. As mentioned by more than ten percent of all subjects

As indicated in Table 7.7 there are no statistically significant differences between experimental conditions with respect to adopted energy conservation behaviors. Most frequently reported energy behavior changes appear to be to use less lighting, lower thermostat one hour before going to bed, and doing the laundry at 60°C. It can be concluded that irrespective of experimental condition subjects in all five groups have adopted a number of energy behavior changes which may be labeled as more attentive natural gas and electricity usage.

Consistent with existing evidence (Ester, 1979b; Ester & van der Meer 1979a + b, 1982; Lipsey, 1977), a factor analysis of energy conservation actions mentioned in Table 7.7 showed that conservation behavior is a cluster of quite heterogeneous behaviors which are not or only weakly interrelated. No clear factors are revealed which justify the construction of sub-clusters. (7)
A subsequent nonparametric experimental groups versus control group analysis (Mann-Whitney U test) did not reveal any statistically significant differences in energy behavior changes between experimental group subjects and control group subjects. Thus, a general trend of more conservation-conscious energy usage can be observed in the whole sample which as such has not been strengthened by experimental manipulations. It appears that subjects have adopted a different number of divergent energy conservation behaviors which resulted in a small but durable conservation effect.

7.6.2 Energy behavior change and specific and general energy attitudes

According to Hypothesis three (H_3) specific attitudes of consumers toward energy conservation are stronger predictors than consumers' general attitudes toward energy scarcity - both as measured before the heating season sets in - of their involvement in specific energy conservation behaviors. Hypothesis five (H_5) stated that consumers' intention to conserve energy - also as measured before the heating season - is a moderate predictor of their involvement in specific energy conservation behaviors in that season (See Section 4.3.1). Both hypotheses will be tested by using the energy behavior changes reported in Table 7.7 as the dependent variables. In order to simplify the data presentation, subjects' specific attitudes toward energy conservation will only be related to the composite scores of $\Sigma B \times E$ and $\Sigma NB \times MC$. Considerations with respect to sample size made it necessary to test both hypotheses on the total sample (experimental groups plus control group). (8) Findings are summarized in Table 7.8.

Table 7.8: Correlations Between Energy Behavior Changes*, and Subjects' Specific and General Energy Attitudes and Intention to Conserve Energy

Conservation Behaviors	$\Sigma B \times E$	$\Sigma NB \times MC$	BI	GCES	Explained Variance
- doing the laundry at 60°C	.10	.09	.06	.07	2%

(Table 7.8 continued)

Conservation Behaviors	$\Sigma B \times E$	$\Sigma NB \times MC$	BI	GCES	Explained Variance
- closing curtains in living-room	.06	.10	.08	.02	1%
- checking thermostat setting	.07	.19	.16	.02	5%
- no pre-washing	.16	.16	.15	.09	5%
- using less lighting	.05	.06	.06	-.01	1%
- closing door between hall and living room	.16	.07	.15	.05	4%
- lowering thermostat one hour before going to bed	.06	.08	.14	-.08	4%
- installing lower-wattage bulbs	.02	.10	.06	.03	2%
- lowering temperature in living room	.03	.09	.06	.01	2%
- lowering thermostat when leaving the house	.06	.10	.09	.01	2%

Note. Intentionally adopted energy conservation behaviors as mentioned by more than ten percent of all subjects.

The findings reported in Table 7.8 clearly show that both specific and general attitudes are hardly correlated with involvement in energy behavior change. At best only 5% of the variance in adopted energy behavior change could be explained by specific energy attitudes ($\Sigma B \times E$, $\Sigma NB \times MC$), the intention to conserve energy (BI), and subjects' general concern with energy scarcity (GCES). Although there is a trend that specific energy attitudes are somewhat "stronger" correlated with energy behavior change than are general attitudes, the data do not confirm H_3. In accordance with H_5, Table 7.8 shows that subjects' pre-experimental intention to conserve energy is only a moderate predictor of their involvement in energy conservation behavior.
The explanations offered in Section 6.2.5 for the observed low explanatory power of specific energy attitudes in predicting behavioral intentions to conserve energy may also apply here. First of all, it could be that salient personal and normative beliefs have not been fully covered in this study. Next, a more detailed inclusion of situational variables and constraints as well as of subjects' past behavioral experience with household energy conservation might

improve the prediction of energy behavior change (Macey & Brown, 1983; Stern, Black & Elworth, 1982, 1983). Also, separate measurements of the different components of the Fishbein model for each conservation behavior should supposedly yield better attitude-behavior predictions (Kok, 1981; Ritsema, Midden and van der Heijden, 1982), though this evokes practical difficulties. Finally, the relatively large time span between the measurements of, respectively, energy attitudes and involvement in energy behavior change may account for low attitude-behavior correlations (Ajzen & Fishbein, 1980, Fishbein & Ajzen, 1975).
The conclusion must be that measurement of energy attitudes with this version of the Fishbein model does not yield stronger attitude-behavior relationship than observed in mainstream social energy research (Anderson & Lipsey, 1977; Cunningham & Lopreato, 1977; Olsen, 1981; Perlman & Warren, 1977; Verhallen & van Raaij, 1981).

7.7 Demographics

For segmentation purposes (e.g. target group definition, design of energy conservation communications) it is important to know whether effectiveness of behavioral interventions aimed at promoting consumer energy conservation is related to demographic characteristics of consumers. This section will briefly look at the relationship between subjects' relative 1980-81 energy consumption and their age, education and socioeconomic status. Since the experimental conditions did not cause any significant effects for electricity consumption, the analysis will be once again restricted to natural gas consumption. Findings are summarized in Table 7.9 (see page 153)

Table 7.9 demonstrates that subjects' relative 1980-81 natural gas consumption in each experimental condition is hardly related to sociodemographic characteristics. No clear overall pattern is revealed which could legitimately suggest explicit sociodemographic criteria for energy conservation segmentation purposes. In most cases there are no statistically significant differences in experimental results in terms of subjects' age, education or socioeconomic status. There are some exceptions, however. It appears, that older subjects decreased their relative 1980-81 natural gas consumption somewhat less as a result of being exposed to conservation information and self-monitoring than younger subjects. Finally, higher educated subjects took somewhat more advantage of conservation information than less educated subjects. (9) Again, it has to be emphasized that those findings are not consistent over experimental conditions.

Table 7.9: Correlations Between Subjects' Relative 1980-81 Natural Gas Consumption and Sociodemographic Characteristics by Experimental Condition

Experimental Condition	Age	Education	Socioeconomic Status	N
	Relative 1980-81 natural gas consumption			
Conservation information	.22*	-.23*	.01	74
Biweekly feedback	-.07	-.02	.15	77
Monthly feedback	-.04	.02	.08	76
Self-monitoring	.19*	-.03	-.16	73
Control group	.10	.05	.06	74

* $p \leq .05$

7.8 Summary

In this chapter the main experimental results were discussed. It was found that all experimental interventions (conservation information, biweekly feedback, monthly feedback, self-monitoring) resulted in reduced natural gas consumption. No effects were found for electricity usage. H_9, H_{10}, H_{11} and H_{12} are therefore partially confirmed. No significant differences between experimental conditions were observed. Therefore, H_{13} had to be rejected.

In spite of the fact that the experimental interventions tested in this study have only been moderately successful in promoting consumer energy conservation, the amount of cubic meters natural gas "saved" could nevertheless be considerable when taken community- or nation-wide, especially with low-cost interventions like conservation information and self-monitoring.

Next, it was found that only a small fraction of subjects' relative 1980-81 natural gas consumption could be explained by their pre-experimental specific and general energy attitudes. Thus, there is only limited support for H_2 and H_4. Also, little empirical support could be found for the "breeding ground hypothesis" (H_{14}) stating that the effectiveness of experimental interventions is related to subjects' pre-experimental attitudes toward energy conservation. It was observed that if thermostat setting agreements exist in households, specific energy conservation attitudes are stronger corrrelated with subjects' relative 1980-81 natural gas consumption.

In accordance with H5 it was found that subjects' pre-experimental behavioral intention to conserve energy is a moderate predictor of their adoption of specific energy conserving behaviors. Next, it was concluded that subjects' pre-experimental attitudes toward energy conservation and energy scarcity were hardly or not correlated with their involvement in specific energy conservation behaviors, which leaves little or no support for H_3. A number of theoretical explanations was offered for this finding.

Finally, sociodemographic correlates of the effectiveness of experimental interventions were discussed.

Notes

1. For full details see de Boer and Ester (1982).
2. See also Section 1.2.
3. Of course this argument only holds if our findings can be generalized to larger entities, which is somewhat questionable given our research locations.
4. In order to reduce the number of degrees of freedom, the four experimental conditions have been taken together in this analysis.
5. Electricity usage was the dependent variable in this study.
6. Although the validity of this measure can be disputed, it nevertheless indicates an overall preoccupation with energy conservation.
7. For a more successful attempt to cluster patterns of energy behavior see van Raaij, W.F., and Verhallen, Th.M.M., Patterns of residential energy behavior, Journal of Economic Psychology, 1983, 85-106.
8. From a theoretical and methodological point of view it would be better to test both hypotheses on the control group only. Sample size considerations (cf. percentages of energy behavior changes reported in Table 7.6), however, exclude this option.
9. Subsequent analyses did not reveal significant correlations between household energy expenditures (budget share) and effectiveness of experimental treatments (cf. Winkler & Winett, 1982).

8. COGNITIVE ASPECTS OF EXPERIMENTAL INTERVENTIONS

8.1 Introduction

The experimental results presented in the preceding chapter call for a more detailed analysis of the psychological context of the behavioral interventions tested in this study, especially given the fact that one of the main hypotheses (H_{13}) about the relative effectiveness of those interventions had to be rejected. This chapter will further explore this psychological context by investigating a number of cognitive aspects which might be related to the effectiveness of the four behavioral interventions.

First of all, the issue of salience of experimental treatments relative to other possible sources of influence will be addressed (Section 8.2). This issue is of crucial importance in view of the likely multiple flows of energy conservation information and conservation prompts subjects are faced with in everyday life. Next, the hypotheses (H_{15}, H_{16}) will be tested about the relationship between subjects' pre-experimental needs for energy conservation information and energy consumption feedback, and the effectiveness of conservation information and feedback (Section 8.3).

In the following three sections how subjects processed the information they were exposed to in the four experimental conditions will be investigated in some detail by examining their interest, evaluation and understanding of the treatments, in order to trace psychological factors and treatment barriers which may have limited the effectiveness of those treatments (Section 8.4, 8.5 and 8.6). In Section 8.7 some data will be provided about Hypothesis nineteen (H_{19}), which assumed increased energy conservation knowledge as a result of being exposed to the experimental interventions, as well as about Hypothesis seventeen (H_{17}) according to which effectiveness of interventions is related to subjects' pre-experimental knowledge of residential energy matters. Next, whether the experimental treatments yielded energy conservation attitude change as predicted by Hypothesis eighteen (H_{18}) will be investigated (Section 8.8). The relationship between energy conservation and consumer comfort is an important issue within the context of energy and consumer policy. This issue will be addressed in Section 8.9 by testing Hypothesis twenty (H_{20}), which assumed no loss of comfort as a result of being exposed to the experimental treatments. In Section 8.10 some data will be presented about the final hypothesis (H_{21}), which stated that positive evaluations of conservation information, feedback, and self-monitoring are related to subjects' attitudes toward wider implementation of those behavioral interventions. Next, possible demographic correlates of the various cognitive aspects discussed in this chapter will be analyzed

(Section 8.11). Finally, the main findings reported in this chapter will be summarized in Section 8.12.

8.2 Salience of experimental interventions

Salience is an important aspect of the psychological context of energy conservation. The less salient energy conservation motives, the less effective behavioral interventions will be to stimulate energy conservation. Given the fact that the most common responses in survey research to energy conservation motives - besides moral considerations - are current energy prices or expected price increases (Olsen, 1981), salience of energy conservation was assumed in this study to be related to subjectively perceived monetary savings from conservation. Consequently, each experimental condition emphasized those monetary savings (See Section 5.3).

The information provided in the four experimental conditions is likely part of a much wider flow of information subjects are exposed to with respect to energy conservation. The existence of such multiple information flows does raise the question of the salience of the experimental treatments relative to other possible sources of influence. In order to test overall salience of information provided in the different experimental conditions, subjects were asked to indicate whether in the past six months they had any discussions with their husband or children about energy conservation, and if so to mention the main reason for those discussions. In order to avoid the measurement of artifacts, interviewers had been carefully instructed not to give the pre-coded answers. (1)

Findings are summarized in Table 8.1

Table 8.1: Reasons for Household Energy Conservation Discussions by Experimental Condition

Reasons	Information & Self-monitoring Group	Feedback Groups	Control Group
	N = 130	N = 142	N = 70
No discussions	23%	17%	21%
Discussions, reported reasons:			
- yearly energy account	24%	24%	24%
- advance energy payment	21%	24%	17%

(Table 8.1 continued)

Reasons	Information & Self-monitoring Group	Feedback Groups	Control Group
	N = 130	N = 142	N = 70
- energy conservation booklet	9%	10%	4%
- feedback letters	2%	17%	3%
- self-monitoring of energy consumption	8%	5%	7%
- discussions with neighbors	11%	11%	7%
- discussions with family, acquaintances, friends not living in neighborhood	10%	6%	10%
- other reasons	21%	29%	36%
- no special reason	15%	14%	10%

Note. Singles excluded; more than one answer possible.

Table 8.1 shows that most subjects (80%) reported discussions in their household about energy conservation in the last six months. The most frequently mentioned reasons for those discussions concern the yearly energy usage account received from the utility company or an advance (monthly or bimonthly) energy payment. It has to be added that all subjects received this account and advance bill in this six-month period. It is also shown that information (conservation booklet, biweekly and monthly feedback letters, self-monitoring) given in the experimental conditions has not been very salient. Only 20% of the subjects in the four conditions refer to this information as a reason for household discussions about energy conservation. The feedback letters seem to have been most successful in this respect.
Thus, it has to be concluded that within this context of multiple information stimuli, the stimuli provided by the experimental conditions can not been characterized as highly salient.

A different though somewhat related issue is whether the experiment did evoke discussions or did stimulate contacts about the experiment between subjects living in the same research location. Such discussions and contacts could have positive effects on the experimental results, but could also confound the findings as far as control group subjects are concerned. Discussions and contacts between neighbors were neither encouraged nor discouraged. However, it was felt to be important to trace in the post-experimental interview whether such discussions and contacts had occurred. Findings are shown in Table 8.2.

Table 8.2: Contacts Between Subjects Living in the Same Research Location About the Experiment by Experimental Condition

Nature of Contacts	Information Group	Biweekly Feedback Group	Monthly Feedback Group	Self-monitoring Group	Control Group
	N = 69	*N* = 74	*N* = 71	*N* = 65	*N* = 70
Spoken about experiment	38%	53%	45%	45%	29%
Received conservation tips	7%	5%	4%	9%	6%
Seen conservation booklet	4%	5%	4%	3%	3%
Seen feedback letters	4%	3%	9%	3%	3%
Seen recording charts	1%	3%	1%	2%	1%

In general it appears from Table 8.2 that some contacts have occurred between subjects, though obviously on a limited scale. As could be expected a priori, experimental subjects discussed the experiment more often than control group subjects. No statistically significant relationship exists between contacts by control group subjects and their relative 1980-81 natural gas consumption. Only small percentages of subjects received conservation tips or exchanged research material (conservation booklet, feedback letters, recording charts). Thus, it may be concluded that the experiment did not produce intensive contacts between subjects about experimental stimuli, which were not aimed at either in this study.

8.3 Pre-experimental needs for conservation information, feedback and self-monitoring, and experimental results

Hypothesis fifteen (H_{15}) and sixteen (H_{16}) stated that the more intense consumers' pre-experimental need for energy conservation information and feedback, the more effective information and feedback are, respectively, in reducing their energy consumption (See Section 4.3.4). (2) These hypotheses were put into the context of the so-called "breeding ground theory", indicating that the effectiveness of experimental interventions will be related to favorable pre-experimental attitudes toward those interventions. Both hypotheses are tested in this section. Besides its theoretical value, they are of interest for possible segmentation purposes.

In Section 6.4 four items were mentioned measuring subjects' pre-experimental need for conservation information and biweekly and monthly feedback. Scores on those items have been transformed into a composite index (α = .66) which will be used in this section. This index is called "perceived effectiveness of conservation information and feedback". Table 8.3 shows the relationship between subjects' scores on this index and their relative 1980-81 natural gas consumption.

Table 8.3: Perceived Effectiveness of Conservation Information and Feedback, and Subjects' Relative 1980-81 Natural Gas Consumption by Experimental Condition

	Information Group	Biweekly Feedback Group	Monthly Feedback Group	Self-monitoring Group	Control Group
	N = 77	N = 77	N = 76	N = 77	N = 76
Perceived effectiveness of conservation information and feedback	-.20*	-.09	-.09	.02	.09

* $p < .05$

Table 8.3 shows very little empirical evidence for H_{15} and H_{16}. Only for the information group is a statistically significant negative correlation found for the relationship between perceived effectiveness of conservation information and feedback and subjects' relative 1980-81 natural gas consumption. For the other experimental groups the hypotheses are not confirmed as the effectiveness of biweekly and monthly feedback is not related to subjects' pre-experimental needs for conservation information and regular feedback.

Probably, the unfamiliarity of subjects with energy consumption feedback in the pre-experimental situation and possible interpretation difficulties may account for these findings. Subsequent sections of this chapter will further explore the psychological context of the behavioral interventions tested in this study, in order to get a more detailed picture of ways in which subjects processed the information they were exposed to.

8.4 Cognitive aspects of conservation information

Consumers interested in energy conservation have a number of possibilities to acquire relevant information. For instance, they may:

a. apply their own conservation knowledge
b. use knowledge of other household members
c. consult information provided by outside sources
d. search for new information

Therefore, it is important to realize - as already argued in Section 8.2 - that the energy conservation booklet given to subjects in this experiment represents only one possibility to acquire energy conservation information.

In this section how subjects processed the conservation information to which they were exposed is analyzed in more detail. First, their interest in and evaluation of the conservation booklet will be investigated (Section 8.4.1), and next, these subjective parameters will be related to subjects' energy consumption (Section 8.4.2).

8.4.1 Energy conservation booklet: interest and evaluation

A number of questions in the post-experimental questionnaire was directed at the energy conservation booklet which subjects received in the four experimental conditions. Issues included: availability, reading intensity, consulting frequency, ascribed effectiveness, and household penetration of the booklet. Results are summarized in Table 8.4.

Table 8.4: Subjects' Interest in and Evaluation of Energy Conservation Booklet

Parameters	Information Group	Feedback Groups & Self-monitoring Group
	N = 69	N = 210
Availability		
Booklet had been kept	74%	73%
Booklet has not been kept	14%	13%
Don't know	3%	11%
Did not receive booklet	9%	2%
Reading intensity		
Read booklet completely	49%	50%
Read booklet partly	23%	22%
Read booklet cursorily	14%	16%
Did not read booklet	4%	10%
Did not receive booklet	9%	2%
Consulting frequency		
Consulted booklet only once	42%	45%
Consulted booklet several times	42%	41%
Don't know	3%	2%
Did not receive booklet/did not read booklet	13%	12%
Ascribed effectiveness		
Helped a lot	3%*	6%*
Helped somewhat	46%	53%
Did not help	35%	25%
Don't know	3%	4%
Did not receive booklet/did not read booklet	13%	12%
Household penetration		
Booklet has been read by others	57%	59%
Booklet has not been read by others	29%	38%
No information	4%	1%
Did not receive booklet	9%	1%

* Wilcoxon's test, $z = 1.82$, $p = .03$ (exclusive of don't know/did not read booklet/did not receive booklet categories)

Table 8.4 clearly shows that with respect to most parameters (availability, reading intensity, consulting frequency, household penetration) there are no statistically significant differences between subjects from the conservation information group and the other experimental groups. Only with regard to ascribed effectiveness of the booklet significant differences are found, indica-

ting that subjects from the feedback and self-monitoring groups attribute greater conservation effectiveness to the booklet compared to subjects from the conservation information group.
Most subjects (73%) indicate they have kept the booklet, whereas 4% reports not to have received or seen it. As far as reading intensity is concerned, it is found that 50% has read the booklet completely, 22% partly, and 15% had read the booklet cursorily. Thus, the booklet has reached about three-quarters of the target group. About 45% has consulted the booklet only once, and another 40% reports to have consulted the booklet several times, often (20%) without a specific reason or for divergent reasons (21%). Some 50% believes the booklet was very or somewhat helpful for energy conservation purposes. If the booklet was felt to have been not helpful (27%), this was in most cases explained in terms of specific circumstances and was not related to the booklet. It can also be observed that the penetration rate of the conservation booklet has not been limited to the target group only, as about 60% of all subjects indicates that the booklet has been read by other household members.

In summary, it can be concluded that in general most subjects have evaluated the conservation booklet quite positively. This conclusion is reflected in the finding that 76% of all subjects believes it to be a good cause to distribute the booklet on a large scale. There appears to be some consistency in this overall judgement given the positive correlations between the parameters mentioned in Table 8.4.
For further analyses a composite score has been computed indicating subjects' overall evaluation of the conservation booklet, including the following parameters: availability of booklet, ascribed effectiveness, large scale distribution of booklet (α = .44, n = 245).
Finally, subjects have also been asked whether they or other household members have searched for energy conservation information in the last six months. Findings show that control group subjects have looked more often for conservation information than experimental group subjects (30% against 18%). This difference could suggest that for a particular segment of the target group the conservation booklet has supplied a certain information need.
Thus, although the conservation information provided to subjects has not been very salient (See Section 8.2), the conservation booklet has been received with interest and appreciation.

8.4.2 Effectiveness of energy conservation booklet

In this section some data will be provided about the relationship between subjects' interest in and evaluation of the energy conservation booklet and the observed effectiveness of this booklet in stimulating household energy conservation. The data analysis will be restricted to information group subjects (N = 69) only. Findings are summarized in Table 8.5.

Table 8.5: Correlation Between Subjects' Interest in and Evaluation of Energy Conservation Booklet and Effectiveness of Conservation Booklet

Consumption Measures	Interest in Conservation Booklet	Evaluation of Conservation Booklet
	_N_ = 69	
Natural gas consumption		
Weighted 1980-consumption	-.10	-.13
Weighted 1981-experimental consumption	-.19	-.19
Relative 1980-81 experimental consumption	-.23*	-.21*
Relative 1980-81 post-experimental consumption	-.03	-.10
Electricity consumption		
Weighted 1980-consumption	-.02	.08
Weighted 1981-experimental consumption	-.03	.08
Relative 1980-81 experimental consumption	-.09	.01
Relative 1980-81 post-experimental consumption	.04	-.03

* $p = .045$

Table 8.5 shows that as far as natural gas consumption is concerned, the effectiveness of conservation information is positively related to subjects' interest in and evaluation of the conservation booklet. However, this relationship only holds for the experimental period. Also, a nonsignificant trend can be observed that subjects with lower weighted experimental natural gas consumption expressed a greater interest in and show a more positive evaluation of the conservation booklet. No statistically significant correlations are found for electricity.

Thus, there appears to be some evidence that the effectiveness of the conservation booklet is related to subjects' appraisal and evaluation of this booklet.

8.5 Cognitive aspects of feedback

One of the main results of this study is that, contrary to the hypothesis, providing subjects with biweekly or monthly energy consumption feedback did not differ in effectiveness from simply supplying subjects with energy conservation information. This unexpected finding calls, of course, for an adequate explanation which will be searched for in cognitive factors affecting feedback. In this respect one may raise the valid question whether in this experiment energy consumption feedback worked according to psychological principles underlying this behavioral intervention. These psychological principles assume that subjects process the feedback information in such a way that they orient themselves on behavioral outcomes by testing those outcomes against implicit or explicit criteria (cf. Ellis & Gaskell, 1978). If energy consumption feedback does not yield intended effects an explanation may be sought for with each of these elements: low salience of feedback information, defective orientation, wrong interpretation of behavioral outcomes, or with testing those outcomes against flexible criteria.

In general, the informational value of feedback depends on the degree in which psychologically relevant factors are made visible and psychologically irrelevant factors are eliminated through correction measures. Examples of psychologically irrelevant factors are: outside temperature, sun and wind orientation, seasonal influences, and caloric value of natural gas. Besides these factors, household energy consumption is also influenced by individual circumstances like holidays, absence patterns, working-hours, rebuildings. Obviously, it is impossible to correct the feedback information for all those factors and circumstances. Apart for corrections made in this study for outside temperature and seasonal influences, subjects had to interpret and subjectively weigh the feedback information themselves. One option for them in this respect is to make consumption comparisons between various feedback periods. However, this option is facilitated by high feedback frequencies which for policy reasons have been deliberately excluded in this experiment.

In summary, the effectiveness of feedback will be related to actively processing feedback information by subjects which involves certain demands with respect to salience and understanding. This section will include a discussion of how subjects processed the feedback information by reporting some data about their evaluation of this information (Section 8.5.1), their understanding of this information (Section 8.5.2), as well as by relating the effectiveness of feedback to subjectively perceived success of conservation efforts (Section 8.5.3).

8.5.1 Biweekly and monthly energy consumption feedback: interest and evaluation

A number of questions in the post-experimental questionnaire was related to subjects' interest in and evaluation of the feedback information they received during the experimental period on trends in their energy consumption. Issues included: availability of feedback letters, reading intensity, between-period comparisons, discussions within household about feedback information, opinion about continuation of feedback, and ascribed effectiveness of regular energy consumption feedback. Results are shown in Table 8.6.

Table 8.6: Subjects' Interest in and Evaluation of Energy Consumption Feedback

Parameters	Biweekly Feedback Group	Monthly Feedback Group
	N = 74	*N* = 71
Availability		
Have kept all feedback letters	70%	75%
Have kept some feedback letters	8%	10%
No feedback letters kept	22%	15%
Reading intensity		
Have read all feedback letters	97%	92%
Have read most feedback letters	1%	7%
Have read some feedback letters	-	1%
Never read feedback letters	1%	-
Between-period comparisons		
Made complete survey	18%	20%
Just made comparisons	55%	55%
Made no comparisons	27%	25%
Household discussions of feedback letters		
Feedback letters have been discussed regularly	49%	46%
Feedback letters have sometimes been dicussed	22%	23%
No discussion of feedback letters	29%	29%
Opinion about continuation of feedback		
Very meaningful	43%	35%
Somewhat meaningful	26%	20%
Not very meaningful	28%	44%
Don't know	3%	1%

(Table 8.6 continued)

Parameters	Biweekly Feedback Group	Monthly Feedback Group
	N = 74	N = 71
Ascribed effectiveness		
Helped a lot	31%*	15%*
Helped somewhat	51%	55%
Did not help	14%	28%
Don't know	3%	1%

* Wilcoxon's test, $z = 2.42$, $p = .008$ after combining did not help/don't know categories

Table 8.6 shows that with respect to most parameters (availability, reading intensity, between-period comparisons, opinion about continuation of feedback, household discussions of feedback letters) there are no statistically significant differences between subjects from both feedback groups. It appears, however, that biweekly feedback group subjects attribute greater effectiveness to feedback in stimulating energy conservation than monthly feedback group subjects. Next, it can be concluded that almost all subjects (94%) have read all feedback letters, though a smaller majority (72%) reports to have kept all feedback letters. Relatively few subjects (19%) have made a complete survey of all feedback letters, most subjects (55%) just made between-period comparisons. In accordance with the earlier observed limited salience of the experimental conditions it is found that about 50% of the subjects report discussions with other household members about the feedback letters. Next, 62% of the subjects believes it to be meaningful to continue with receiving regular feedback information about their energy consumption. Finally, the majority of subjects (76%) indicates that the feedback information helped a lot or somewhat to conserve energy.

Although not included in Table 8.6, subjects' evaluation of feedback concerning the various parameters mentioned in this table show some consistency given positive correlations between those parameters.

For further analyses a composite score has been computed measuring subject's overall evaluation of feedback, including the following parameters: ascribed effectiveness of feedback, opinion about continuation of feedback, and attitude toward wider implementation of feedback ($\alpha = .69$, $N = 143$).

Finally, subjects were asked whether they believed it to be a good cause when utility companies would supply consumers with regular feedback about trends in their energy consumption. It is found that 72% of all subjects would support such an initiative.

Thus, one may conclude that in general subjects have received the feedback letters with interest and evaluate energy consumption feedback in quite positive terms.

8.5.2 Subjective understanding of feedback and effectiveness of feedback

One of the intended effects of energy feedback is that in due time subjects are able to explain decreased or increased trends in their energy consumption by relating those trends to behavioral performances or specific circumstances. In Section 3.2.3 this process was referred to as the learning function of feedback. In order to see whether those intended effects occurred, subjects were asked to indicate whether they usually could or could not explain the energy consumption trend information contained in the feedback letters. Results are shown in Table 8.7.

Table 8.7: Subjective Understanding of Decreased or Increased Trends in Energy Consumption as Indicated in Feedback Letters

	Biweekly Feedback Group	Monthly Feedback Group
	N = 74	N = 71
Natural gas consumption trends		
Could usually explain trends	46%	43%
Could sometimes explain trends	18%	14%
Could usually not explain trends	36%	43%
Electricity consumption trends		
Could usually explain trends	51%	44%
Could sometimes explain trends	8%	9%
Could usually not explain trends	41%	48%

As Table 8.7 demonstrates, more than 40% of all subjects could usually explain decreased or increased trends in their natural gas and electricity consumption as shown in the feedback letters, but also more than 40% could usually not explain the feedback information they received. No differences in this respect are found between natural gas and electricity consumption. Thus, apparently subjects have mixed feelings about understanding energy consumption feedback information. Subsequent analyses revealed that subjective understanding of feedback is related to interest in and evaluation of the conservation booklet and of feedback.

Subjective understanding of feedback shows positive relationships with reading intensity of the conservation booklet (R = .32, N = 142), with having made between-period comparisons (R = .24, N = 141), and with overall evaluation of feedback (R = .30, N = 141). There is also evidence for a positive relationship between subjects' overall evaluation of the conservation booklet and their evaluation of feedback (R = .31, N = 129).

Thus, from a subjective point of view there appears to be empirical support for the central role played by individual comprehensibility of energy consumption feedback information. However, relating these variables to real consumption trends cannot confirm the significance of this role, as shown in Table 8.8.

Table 8.8: Correlations Between Subjects' Interest in Evaluation and Understanding of Feedback, and Effectiveness of Feedback

Consumption Measures	Interest in Feedback	Evaluation of Feedback	Understanding of Feedback
	$\underline{N}$ = 145		
<u>Natural gas consumption</u>			
Weighted 1980-consumption	.02	-.01	.03
Weighted 1981-experimental consumption	-.01	-.09	-.01
Relative 1980-81 experimental consumption	-.06	-.21*	-.06
Relative 1980-81 post-experimental consumption	-.03	.08	-.03
<u>Electricity consumption</u>			
Weighted 1980-consumption	-.02	.08	-.08
Weighted 1981-experimental consumption	-.01	.09	-.06
Relative 1980-81 experimental consumption	.02	.04	.02
Relative 1980-81 post-experimental consumption	-.02	.08	-.01

* $\underline{p}$ = .05

Table 8.8 clearly demonstrates that - with one exception - effectiveness of feedback is not related to subjects' interest in, evaluation and understanding of feedback. Only in case of natural gas consumption effectiveness of feedback appears to be highest among subjects with a positive overall evaluation of feedback. This finding, however, is not extended to the post-experimental period.

Thus, on the level of real consumption measures the role of cognitive factors subjectively associated with feedback cannot be substantiated.

8.5.3 Perceived success of conservation efforts and effectiveness of feedback

In the previous two sections the issue was addressed of how subjects processed the feedback information. This section will turn to the related issue of whether feedback performed a confirmatory function. This confirmatory function is especially of importance to subjects who consciously changed their behavior in order to conserve energy. In their case feedback may reinforce energy behavior changes by confirming its effectiveness and, thus, by motivating those subjects to continue their conservation efforts.

As described in Section 7.6 subjects were asked in the post-experimental questionnaire to select from a set of 31 different energy-related behaviors those behaviors they had intentionally adopted for energy conservation purposes. In addition to this question, subjects who reported energy conservation behavior changes were next asked to indicate whether they perceived those efforts as being successful. Results are shown in Table 8.9.

Table 8.9: Perceived Success of Conservation Efforts by Experimental Condition

Perceived Success	Conservation Information Group	Feedback Groups	Self-monitoring Group	Control Group
	N = 48	N = 103	N = 42	N = 47
Natural gas consumption				
Very successful	11%[a]	20%[b]	10%[c]	9%[d]
Somewhat successful	36%	45%	41%	28%
Not successful	13%	20%	12%	26%
Don't know	40%	16%	37%	38%

(Table 8.9 continued)

Perceived Success	Conservation Information Group	Feedback Groups	Self-monitoring Group	Control Group
	N = 48	*N* = 103	*N* = 42	*N* = 47
Electricity consumption				
Very successful	6%	9%	7%	6%
Somewhat successful	35%	34%	19%	17%
Not successful	21%	37%	31%	32%
Don't know	38%	20%	43%	45%

a + b + c against d: Wilcoxon's test $z = 2.48$, $p = .007$.
a against b : Wilcoxon's test $z = 1.47$, $p = .07$.

As can be concluded from Table 8.9, about 65% of feedback group subjects perceive their conservation efforts with respect to natural gas consumption as being successful, and 43% with respect to electricity consumption. No statistically significant differences are found between both feedback groups. Both conservation success perceptions tend to be interrelated ($R = .45$, $p = .01$). As far as natural gas is concerned, experimental group subjects report greater conservation success than control group subjects, which as such reflect the general experimental outcomes (See Section 7.4). No statistically significant differences are observed for electricity conservation efforts. With respect to natural gas conservation efforts a non-significant ($p = .07$) tendency is found for greater conservation success in both feedback conditions compared to the conservation information condition. The main difference, however, is the percentage of don't know answers in both groups (16% against 40%).

Finally, perceived success of conservation efforts is related to subjects' relative 1980-81 natural gas consumption. It has to be emphasized that, as mentioned before (See Chapter 7), subjects' mean natural gas consumption in the total sample shows a decreasing trend, which indicates that a relative consumption score of .00 already signifies some conservation. Results are summarized in Table 8.10.

Table 8.10: Consumption Scores Broken Down by Perceived Success of Conservation Efforts and Subjects' Relative 1980-81 Natural Gas Consumption

	Conservation Information Group		Feedback Groups		Self-monitoring Group		Control Group	
Natural Gas	x̄	(N)	x̄	(N)	x̄	(N)	x̄	(N)
Very successful	+.01	5	-.07	20	-.03	4	-.13	4
Somewhat successful	-.03	17	-.01	46	-.02	17	.00	13
Not successful	-.01	6	+.03	20	.00	5	+.09	12
Don't know	-.01	19	.00	16	-.03	15	+.04	18
Total	-.01	47	-.01	102	-.02	41	+.03	47

Note. By ln-transformation the consumption scores equal proportions increase or decrease with respect to mean consumption trend of total sample

Obviously, the most remarkable finding shown in Table 8.10 are the relatively extreme control group scores. If much success is attributed to conservation efforts, their relative natural gas consumption is quite below the mean consumption trend, whereas the opposite is true when low success of conservation efforts is reported. Given small cell size, some caution is needed in interpreting these findings.
The confirmatory function of feedback would lead one to expect that perceived success of conservation efforts in both feedback groups is stronger related to subjects' relative 1980-81 natural gas consumption than in the two other experimental groups. Although, a non-significant (p = .07) trend is observed for differences between experimental groups and control group, there appear to be no statistically significant differences between experimental groups. Thus, one is forced to conclude that energy consumption feedback did not provide subjects with confirmatory information about behavioral performance.

In summary, although biweekly and monthly energy consumption feedback has been received with interest and is generally evaluated in quite positive terms, it could not been demonstrated that subjects processed the feedback information in accordance with psychological principles underlying feedback. It was found that a significant segment of feedback group subjects experienced difficulties in understanding the feedback information, however, no emprirical support could be observed for a positive relationship between interest for feedback, comprehensibility of feedback information, and conservation trends. Finally, feedback did not supply subjects with confirmatory evidence about behavioral performance. Therefore, one has to conclude that biweekly and monthly energy consumption feedback frequencies, chosen for policy reasons, yield much less effective results than studies which used daily feedback frequencies (See Chapter 3). As such there is no empirical evidence which could legitimately support large scale implementation of biweekly or monthly energy consumption feedback to consumers, at least not as provided in this study.

8.6 Cognitive aspects of self-monitoring

Self-monitoring is based on a psychological process through which subjects observe their own behavioral performance by testing this performance against implicit or explicit criteria. As outlined in Section 3.2.4 regular self-monitoring of residential energy consumption - through frequent meter readings - may have at least four functions: attention, orientation, discussion, and confirmation. In this section the psychological context of self-monitoring will be further explored. First, some general facts about subjects' response to the recording prompt will be presented (Section 8.6.1), as well as some findings with respect to their interest in and evaluation of self-monitoring of household energy consumption (Section 8.6.2). Next, two additional cognitive factors possibly affecting the effectiveness of this experimental intervention are discussed: subjects' understanding of self-monitoring procedures (Section 8.6.3), and subjects' perceived effectiveness of conservation efforts (Section 8.6.4).

8.6.1 Some facts about response to self-monitoring request

Table 8.11 contains some general data about subjects' response to the recording prompt in the self-monitoring condition.

Table 8.11: Some Facts About Self-Monitoring Activities

Parameters		
Response to self-monitoring request		N = 77
Recording chart has been returned		45%
Recording chart has not been returned but self-recording activities are reported		7%
Recording chart has not been returned		48%
Recording period		
	Subjects (N = 77)	Median of readings
Half January - Early March	30%	9
Early March - End of April	38%	24
End of April - June	27%	25
Number of recording periods per subject		N = 35
One period		29%
Two periods		34%
Three periods		37%
First month of recording		N = 35
January		60%
February		6%
March		34%
Quality indication of recording		N = 35
Computation of energy consumption plus temperature recording		77%
Computation of energy consumption, but no temperature recording		9%
No computation of energy consumption		14%
Maximal recording frequency per period		N = 35
Daily* recordings (21 - 42 readings)		51%
Weekly* recordings (6 - 20 readings)		43%
Marginal recordings		6%
Overall recording frequencies		N = 77
Daily* recordings		23%
Weekly* recordings		19%
Marginal** recordings		9%
No recordings		48%

Note. * Approximately. ** Inclusive of subjects which did not sent back recording charts but reported self-monitoring activities.

As shown in Table 8.11 about half (52%) of households assigned to the self-monitoring condition participated in some degree in the requested energy consumption recording during the experimental period. A subsequent analysis revealed that some 30% of subjects assigned to the other conditions (N = 284) report to practice self-monitoring activities. No statistically significant differences are found in this respect between experimental group subjects and control groups subjects.

As mentioned in Section 5.3.3, subjects had been advised to read their energy meters at least with weekly frequencies, though in addition they were informed that more frequent readings would enhance their understanding of household energy usage. In the second and third six-week recording period subjects were instructed to record the meter readings on a daily basis. It can be concluded that this instruction increased daily recording frequencies, as the recording median rose from 9 times (about weekly) to 24 times (about every two days) in the concerned periods.

The number of subjects participating in each recording period differed somewhat. It appears that the recording activities were especially concentrated in the second period (beginning of March to end of April). This finding is of importance, since in this period the very cold days were over.

It is not easy to give an indication of the overall quality of meter recordings by subjects. Some recording charts contained detailed information, including outside temperature, energy consumption calculations, and recording of special circumstances. Other charts, however, contained less detailed information. But the overall impression of recording quality is quite favorable, given the fact that almost 80% of the returned charts contained rather detailed recordings.

Next, it can be observed that for those subjects which complied with the self-monitoring request, the maximal recording frequency per six-week period is about daily (51%) or weekly (43%). Finally, it can be concluded from Table 8.11 that of all self-monitoring group subjects, with respect to the entire experimental period, 23% of all subjects recorded their energy consumption with daily frequencies, 9% only recorded marginally, whereas 48% did not comply with the self-monitoring request.

Thus, it appears that a relatively simple behavioral intervention like frequent energy consumption recording is quite appealing to consumers and can be rather easily initiated by asking them to participate (cf. Winett, Neale & Grier, 1979).

8.6.2 Self-monitoring of energy usage: interest and evaluation

The main criterion to measure subjects' interest for self-monitoring of energy usage is of course the number of subjects which complied with the recording request, whereas evaluation of self-monitoring may be related to recording frequency. However, the post-experimental questionnaire contained a number of additional questions with respect to self-monitoring which may further complete our understanding of subjects' interest in and evaluation of this behavior

change technique.
First of all, it was asked which household member recorded the energy meter readings. Findings are shown in Table 8.12.

Table 8.12: Meter Reading Recordings by Household Member

Recording Person	Self-monitoring Group	Other Conditions
	N = 34	N = 86
Housewife	56%	40%
Partner	26%	33%
Other household member	6%	12%
No fixed person	12%	16%

As demonstrated in Table 8.12 self-monitoring of residential energy usage is predominantly performed by a fixed household member, which in this experiment usually was the housewife. (3) In the other conditions there are hardly pronounced differences in recording activities between housewife and partner.
Next, a number of questions was asked related to subjects' interest in and evaluation of self-monitoring. Issues included: reasons for nonparticipation, attitude toward self-monitoring among non-participants, preferred recording frequency, household discussions of consumption figures, ascribed effectiveness of self-monitoring, and opinion about continuation of self-monitoring. Findings are summarized in Table 8.13. In view of the small number of subjects involved, those findings have to be interpreted carefully.

Table 8.13: Subjects' Interest in and Evaluation of Self-Monitoring of Energy Usage

Parameters	Self-monitoring Group
Reasons for nonparticipation	N = 25
Too complicated	4%
Too cumbersome	25%

(Table 8.13 continued)

Parameters	Self-monitoring Group
No time	29%
Conserve enough energy already	7%
Other reasons	36%
Attitude toward self-monitoring among nonparticipants	N = 25
Could be helpful	16%
Could be somewhat helpful	52%
Could not be helpful	32%
Preferred recording frequency	N = 39
Daily	44%
Every few days	18%
Weekly	28%
Every few weeks	-
Monthly	8%
Otherwise	3%
Household discussions of consumption figures	N = 37
Yes, consumption figures have been discussed	54%
No, consumption figures have not been discussed	46%
Ascribed effectiveness of self-monitoring	N = 39
Helped a lot	21%
Helped somewhat	41%
Did not help	36%
Don't know	3%
Opinion about continuation of self-monitoring	N = 39
Very meaningful	41%
Somewhat meaningful	18%
Not very meaningful	36%
Don't know	5%

As can be seen in Table 8.13 the reasons for nonparticipation are somewhat diverse but seem to be especially related to time considerations (29%) and expected intricacy (29%) of recording activities. The attitude toward self-monitoring among nonparticipants, however, is rather favorable as 68% feels that self-monitoring could be somewhat or very helpful in stimulating household energy conservation. Next, daily (44%) and weekly (28%) recording frequencies are preferred by participating subjects over less frequent recordings. In slightly more than 50% of the households involved the consumption figures have been discussed with other household members. The majority of subjects (62%) attribute some or very much effectiveness to self-monitoring in encouraging household energy conservation, whereas 26% holds opposite views. Finally, about 60% of participating subjects feels that continuation of self-monitoring would be meaningful.

Further analyses indicated some marked differences with repect to self-monitoring between subjects living in insulated dwellings and subjects living in non-insulated dwellings. Table 8.14 contains some of those differences.

Table 8.14: Self-Monitoring and Insulation of Dwellings

	Insulated Dwellings	Non-insulated Dwellings
Self-monitoring condition	N = 33	N = 44
Daily recordings	42%[a]	9%[b]
Weekly recordings	21%	18%
Marginal recordings	9%	9%
No recordings	27%	64%
Other conditions	N = 140	N = 144
Reported self-monitoring activities	35%[c]	26%[d]
No self-monitoring activities reported	65%	74%

a against b: Wilcoxon's test $z = 3.64$, p .001

c against d: Wilcoxon's test $z = 1.58$, $p = .06$

As can be observed from Table 8.14 subjects living in insulated dwellings have complied significantly more with the request to monitor their household energy usage than subjects living in non-insulated dwellings (63% against 27%). Also, they recorded their energy meter readings with higher frequencies. In the other conditions a non-significant trend is found for more self-monitoring activities among residents of insulated dwellings. One may hypothesize that subjects who have invested in insulation of their dwellings or who pay a rent increase for insulation are more interested in observing trends in their household energy consumption. Their greater involvement in this respect may explain why it was easier to activate them to comply with the self-monitoring request. In view of segmentation purposes residents of insulated dwellings could therefore be an interesting target group for self-monitoring campaigns.

8.6.3 Subjective understanding of self-monitoring

An important cognitive aspect of the psychological context of self-monitoring is that subjects must be able to explain observed trends in their household energy usage by relating those trends to certain (changes in) patterns of energy-related household behaviors. Table 8.15 contains some information about subjects' understanding of decreased or increased energy consumption trends as made visible by frequent recordings of their energy meters.

Table 8.15: Subjective Understanding of Decreased or Increased Trends in Energy Consumption Through Self-Monitoring of Energy Usage

	Natural Gas Consumption Trends	Electricity Consumption Trends
	N = 39	N = 39
Could usually explain trends	72%	62%
Could sometimes explain trends	3%	10%
Could usually not explain trends	18%	15%
Don't know	8%	13%

As Table 8.15 demonstrates, subjects could usually explain decreased or increased energy consumption trends as indicated by the self-monitoring activities. There are no pronounced differences between understanding observed natural gas and electricity consumption trends. Comparing those findings with subjects' understanding of consumption trends as indicated in the feedback letters (See Table 8.7), it appears that greater understanding of consumption trends is reported by self-monitoring subjects. This result could be explained by the fact that subjects were more actively involved in the self-monitoring condition compared with subjects in both feedback conditions.

8.6.4 Perceived success of conservation efforts and effectiveness of self-monitoring

Subjects who actively complied with the self-monitoring request appear to be already relatively low energy users given their baseline energy consumption.

Accounting for insulation of dwellings, it is found that subjects who recorded their energy meter readings with at least weekly frequencies are characterized by a weighted 1980 natural gas consumption which is about 8% lower than subjects who did not or only marginally monitor their energy usage. This pattern remains unchanged, however, during the period of experimental intervention, which may be explained by the fact that the main recording activities were concentrated in the second recording period which had relatively few cold days (See Section 8.6.1). As far as post-experimental consumption differences are concerned, one must account for a general trend difference between subjects living in insulated or non-insulated dwellings. Since insulation and self-monitoring appeared to be related, a correction factor was computed based on observed relationships in the other conditions. Given this adjusted weighted 1980 natural gas consumption a trend is found that subjects who recorded their energy usage with about daily frequencies have significantly lower post-experimental consumption levels than other subjects in the self-monitoring condition. Again, this difference was only observed for natural gas consumption and not for electricity consumption. Thus, there are some indications which demonstrate the effectiveness of intensive self-monitoring of energy usage.

Finally, it will be analyzed whether self-monitoring performed a confirmatory function by applying the same analysis as with respect to feedback (See Section 8.5.3). In view of the small number of subjects, it was decided to extend the analysis to all subjects who report involvement in self-monitoring activities. An additional advantage is that a distinction can be made between subjects who do the recordings themselves and those with other family members doing the recordings.

Table 8.16 shows relationships between recording activities and perceived success of conservation efforts (See also Table 8.9).

Table 8.16: Perceived Success of Conservation Efforts by Self-Monitoring of Energy Usage

	No Recording	Recording Done by Other Family Member	Recording Done by Subject
Natural gas consumption	N = 149	N = 51	N = 37
Very successful	13%[a]	12%	19%[b]
Somewhat successful	38%	29%	59%
Not successful	19%	22%	8%
Don't know	30%	37%	14%

(Table 8.16 continued)

	No Recording	Recording Done by Other Family Member	Recording Done by Subject
Electricity consumption	N = 151	N = 52	N = 37
Very successful	6%	8%	14%
Somewhat successful	28%	25%	32%
Not successful	32%	35%	27%
Don't know	34%	33%	27%

a against b: Wilcoxon's test z = 2.65, p = .004

As demonstrated in Table 8.16, subjects who record their energy usage report significantly more conservation success with respect to natural gas consumption than subjects who do not record their usage. Moreover, this difference only holds if subjects themselves do the recordings. No differences are observed with respect to electricity consumption.

The relationships between perceived success of natural gas conservation efforts, self-monitoring activities and subjects' relative 1980-81 natural gas consumption is shown in Table 8.17.

Table 8.17: Consumption Scores Broken Down by Perceived Success of Conservation Efforts and Subjects' Relative 1980-81 Natural Gas Consumption by Self-Monitoring Activities

	No Recording		Recording Done by Other Family Member		Recording Done by Subject	
Natural Gas	$\bar{x}$	(N)	$\bar{x}$	(N)	$\bar{x}$	(N)
Very successful	-.04	(20)	-.08	(6)	-.12	(7)
Somewhat successful	-.02	(56)	-.03	(15)	.00	(22)
Not successful	+.05	(29)	+.04	(11)	-.06	(3)
Don't know	.00	(44)	.00	(19)	+.04	(5)
Total	.00	(149)	-.01	(51)	-.02	(37)

Note. By ln-transformation the consumption scores equal proportions increase or decrease with respect to mean consumption trend of total sample.

As demonstrated in Table 8.17, the difference observed in Table 8.16 is remained when subjects' relative 1980-81 natural gas consumption is taken into account. If subjects' relative 1980-81 consumption approximately equals that of others, subjects who record their energy usage themselves report greater success of conservation efforts (*p* = .008). Table 8.17 shows that subjects who do the recording activities themselves and who report some conservation success have on an average the same relative consumption levels compared to no-recording subjects who don't know whether their conservation efforts yielded any success. This finding could imply that self-monitoring provided subjects with confirmatory evidence about the success of their conservation efforts, though this conclusion is based on correlational data only.

In summary, there appears to be some positive indications of the effectiveness of frequent self-monitoring of household energy usage. Those indications concern observed conservation trends and confirmation of conservation efforts. In view of the obvious low-cost nature of self-monitoring, this seems an interesting conclusion for energy conservation policies which are willing to include behavioral techniques and interventions.

8.7 Energy knowledge change

In this section two hypotheses will be tested about the role of residential energy knowledge. Hypothesis seventeen (H_{17}) stated that the more accurate subjects' pre-experimental knowledge of residential energy matters, the more effective information, feedback, and self-monitoring are in reducing their energy consumption. Hypothesis nineteen (H_{19}) assumed that if subjects reduce their energy consumption as a result of being exposed to the experimental interventions, their knowledge of residential energy matters will change accordingly by becoming more accurate.

With respect to H_{17}, correlational analyses indicate no statistically significant relationship between subjects' relative 1980-81 natural gas consumption and their pre-experimental energy knowledge both during and after experimental treatment within each experimental group. In the control group, however, relative natural gas consumption is statistically significant negatively correlated with pre-experimental energy knowledge in both periods (R = -.21, *p* = .037, R = -.28, *p* = .007, respectively). A subsequent analysis revealed a significant interaction effect of energy knowledge and experimental results, both during (*p* = .035) and after (*p* = .002) experimental treatment. Next, a BREAKDOWN analysis was done which leads to the following specification of the experimental results: the difference in relative 1980-81 natural gas consumption is absent for subjects with some basic (pre-experimental) energy knowledge and higher for subjects with low energy knowledge levels. In percentual differences:

low energy knowledge - experimental effect : 8%
some energy knowledge - experimental effect : 4%
higher energy knowledge - experimental effect : 0%

Those findings imply that H_{17} is not supported. In fact, the opposite appears to be true: the experimental interventions were most effective among subjects with low knowledge levels of residential energy matters.

In order to test H_{19}, subjects were asked - like in the pre-experimental interview (See Section 6.3) - to indicate the price of one cubic meter gas and one KWH electricity. In addition they were asked to choose in three cases the most effective energy conserving behavior from a given set of two behaviors (e.g. to indicate whether lowering the heating thermostat one degree (C.) during daytime is more or less effective than five degrees at night. The cases selected had all been discussed in the conservation booklet. Table 8.18 shows subjects' answers to those five knowledge questions by experimental condition.

Table 8.18: Energy Knowledge Change by Experimental Condition

Knowledge Items	Conservation Information Group	Feedback & Self-monitoring Groups	Control Group	Pre-Experimental Interview
	$\underline{N}$ = 69	$\underline{N}$ = 210	$\underline{N}$ = 70	$\underline{N}$ = 470
	% correct answers			
Price knowledge natural gas	23%	33%	19%	8%
Price knowledge electricity	14%	23%	9%	8%
Energy knowledge case one	51%	34%	34%	-
Energy knowledge case two	74%	84%	89%	-
Energy knowledge case three	30%	27%	24%	-

Table 8.18 demonstrates that only in case of price knowledge are there significant differences between experimental and control group subjects. However, both

natural gas and electricity prices were indicated by stickers on the self-monitoring recording charts. When excluding self-monitoring subjects, differences in price knowledge between experimental and control group subjects are no longer statistically significant. Thus it cannot be demonstrated that the experimental treatments increased subjects' knowledge of residential energy knowledge. Consequently, there is no empirical support for H_{19} either.
It is noteworthy that in case of natural gas price knowledge, control group subjects show higher knowledge levels compared to the pre-experimental situation (19% versus 8% correct answers). This finding is probably related to the large amount of publicity given to natural gas price increases in early 1981. In the study by Ritsema, Midden, and van der Heijden (1982) a similar percentage (20%) was found of the Dutch adult population that could accurately indicate the price of one cubic meter natural gas.

Further analyses showed that with the exception of the two price items ($R = .62$, $p = .001$) the energy knowledge variables are not interrelated, which points out that subjects did not systematically give correct or incorrect answers. Also, it was found that only in the case of these two items was energy knowledge (weakly) correlated with reading intensity and consulting frequency of the conservation booklet. No significant relationships were observed with overall evaluation of the booklet. Finally, it could not be demonstrated that subjects' energy knowledge is correlated with their relative 1980-81 natural gas use.
Thus, although there is some evidence - contrary to our hypothesis - that the conservation booklet has been most effective for subjects with low pre-experimental knowledge levels, it could not be shown that the booklet has increased energy knowledge.

8.8 Energy attitude change

As argued in Chapter 3 the relationship between effectiveness of behavior change interventions aimed at promoting energy conservation and energy attitude change is an often neglected issue in experimental energy research. Cognitive changes as a possible result of being exposed to experimental interventions have been only scarcely investigated in behavioral energy experiments. This section will present some empirical data about those cognitive changes.
Hypothesis eighteen (H_{18}) stated that if consumers reduce their energy consumption as a result of being exposed to conservation information, feedback, and self-monitoring, their attitudes toward energy conservation will change accordingly by becoming more positive (See Section 4.3.4). In this section an at-

tempt will be made to determine the degree of empirical support for this hypothesis.

Strictly speaking, the most thorough test of H_{18} would have been to repeat the forty-six attitude items - measuring subjects' specific attitudes toward residential energy conservation - from the pre-experimental questionnaire in the post-experimental survey. However, space limits did not permit such an extensive attitude change measurement. Consequently, it had to be decided to use a more direct measurement of energy attitude change (cf. Section 5.8.2). Therefore, subjects were simply asked in the post-experimental interview whether they held more favorable or less favorable attitudes toward energy conservation given their experiences in the past six months. Findings are shown in Table 8.19.

Table 8.19: Energy Conservation Attitude Change by Experimental Condition

Experimental Condition		More Positive Attitudes	No Change in Attitudes	Less Positive Attitudes	Don't know
		% Change			
Conservation information	(_N_ = 69)	55%	36%	7%	1%
Biweekly feedback	(_N_ = 74)	69%	22%	8%	1%
Monthly feedback	(_N_ = 71)	58%	37%	6%	-
Self-monitoring	(_N_ = 69)	55%	34%	9%	2%
Control Group	(_N_ = 70)	53%	39%	9%	-

Obviously, one has to conclude from Table 8.19 that on the level of subjectively perceived energy conservation attitude change, there are no pronounced differences between experimental groups and control group. In <u>all</u> groups more than 50% of the subjects report more positive attitudes toward energy conservation. Attitude change appears to be relatively greatest for biweekly feedback subjects. Within experimental groups there is no significant relationship between energy conservation attitude change and subject's relative 1980-81 natural gas use. Consequently, H_{18} is not supported. (4)

8.9 Comfort experiences

Several authors have convincingly arqued that energy conservation is not necessarily related to loss of comfort (Yergin, 1979). On the contrary, our standard of living and quality of life could very well be maintained when conserving considerable amounts of energy (Potma, 1979). Empirical support for this important conclusion was found in an experimental study by Winett, Hatcher, Fort, Leckliter, Love, Riley, and Fishback (1982) which has been discussed in Section 3.3.2.1 and 4.3.4. Their results showed that consumers could make quite substantial changes in their thermostat settings without experiencing loss of comfort. These findings suggest that no individual sacrifices are involved in adopting many simple conservation practices.
Partly based on the outcomes of the Winett et al. (1982) experiment it was assumed in Hypothesis twenty (H_{20}) that consumers who reduced their energy consumption during the experimental conditions will not differ in experienced thermal comfort from consumers who did not reduce their energy consumption (See Section 4.3.4). This hypothesis will be tested in this section.

First of all, it has to be mentioned that space constraints permitted the inclusion of only a limited number of questions in the post-experimental questionnaire with respect to comfort experiences. These limitations have to be taken into account when interpreting the findings to be reported in this section. In total, only four attitude questions were included which addressed comfort matters. Thus, at best, only marginal attention could be paid in this study to comfort preferences and comfort experiences. Responses to those four attitude items are shown in Table 8.20.

Table 8.20: Comfort Importance and Reported Loss of Comfort: Experimental Groups Versus Control Group

Comfort Items		Experimental Groups	Control Group
		N = 279	N = 70
1. I think that the living-room temperature should be comfortable such that one is not forced to wear a sweater or a vest	agree	58%	60%
	neutral	4%	3%
	disagree	38%	37%
2. We (I) soon suffer from low temperatures	agree	52%	47%
	neutral	7%	11%
	disagree	41%	41%

(Table 8.20 continued)

Comfort Items		Experimental Groups	Control Group
		N = 279	N = 70
3. It would have been bad for my and my family's health if I had turned down the heating thermostat a few degrees this winter	agree neutral disagree	27% 9% 63%	23% 17% 60%
4. It would have been better for comfort in this home if I had raised the heating thermostat a few degrees this winter	agree neutral disagree	13% 6% 80%	13% 13% 74%

As Table 8.20 demonstrates, there are no (statistically significant) differences between experimental group subjects and control group subjects with respect to the four comfort items. About 60% of all subjects feels that the living room temperature should be comfortable such that no extra clothing is required, whereas the same percentage disagrees with the likelihood of negative health effects if they had lowered the heating thermostat a few degrees last winter. Some 50% feels that they soon suffer from low temperatures, but about 80% of the subjects disagrees with the statement that for comfort reasons higher inside temperatures would have been preferable last winter.

In order to relate those comfort items to subjects' relative 1980-81 natural gas consumption, scores on the first three items from Table 8.20 have been transformed into a composite index. Given the divergent nature of these items, the internal consistency of this index is quite reasonable ($\alpha = .43$). This index is called "comfort importance". The remaining item (item four) will be referred to as "reported loss of comfort". It appears, that comfort importance is negatively correlated with concern about energy scarcity ($R = -.16$, $p = .002$), and with intention to conserve energy ($R = -.20$, $p = .001$). No statistically significant relationships were found in this respect for reported loss of comfort.

Table 8.21 contains the correlations between comfort importance, reported loss of comfort, and subjects relative 1980-81 natural gas consumption.

Table 8.21: Relationship Between Comfort Importance, Reported Loss of Comfort and Subjects' Relative 1980-81 Natural Gas Consumption (During and After Experimental Period)

	During Experimental Period		After Experimental Period	
	Experimental Group N = 279	Control Group N = 70	Experimental Groups N = 269	Control Group N = 70
Comfort importance	.08	.05	.17*	.15
Reported loss of comfort	-.14*	-.11	-.05	-.15

* $p = \leq .01$

Although the findings are not very pronounced and conclusive, it is shown in Table 8.21 that during the experimental period subjects' relative 1980-81 natural gas consumption is statistically significant negatively related to reported loss of comfort. This finding is not, however, extended to the post-experimental period. Next, it is found that in the post-experimental period comfort importance is positively correlated to subjects' relative 1980-81 natural gas consumption but not, however, in the experimental period. No statistically significant differences are observed for the control group subjects.

However small those differences, the Hypothesis (H_{20}) stating that exposure to experimental treatments does not result in reported loss of comfort has to be rejected. Further disconfirming evidence is given by the finding of a significant partial correlation between subjects' relative 1980-81 natural gas consumption in the post-experimental period and comfort importance ($R = .16$, $p = .003$) given their weighted 1980 consumption, their weighted experimental consumption and a dummy variable for insulation. This correlation suggests that a post-experimental conservation decline is somewhat related to comfort importance.

Again, it has to be emphasized that comfort issues were only briefly addressed in the post-experimental questionnaire. Nevertheless it has to be concluded that a number of subjects in the experimental groups reported loss of comfort. It would be of interest to separately investigate a number of energy conservation behaviors which differ in energy consequences but are equal with respect to comfort impacts.

8.10 Attitudes toward large-scale implementation of experimental interventions

Hypothesis twenty-one (H_{21}) stated that the more positive consumers' evaluations of energy conservation information, feedback, or self-monitoring, the more positive their attitudes toward wider implementation of these conservation strategies by public utility companies (See Section 4.2.4). This hypothesis will be tested in this section.
First of all, some data are provided about subjects' attitudes toward large-scale implementation of the conservation booklet, feedback, and self-monitoring.

Table 8.22: Attitudes Toward Large-Scale Implementation by Public Utility Companies of Conservation Booklet, Feedback, and Self-Monitoring by Experimental Condition

	Conservation Information Group	Biweekly Feedback Group	Monthly Feedback Group	Self-monitoring Group
	N = 69	N = 74	N = 71	N = 65
Conservation Booklet				
Good cause	77%	73%	82%	74%
Depends	6%	7%	10%	6%
No good cause	6%	8%	3%	9%
Don't know	3%	10%	4%	9%
No information	9%	3%	1%	2%
Feedback				
Good cause		76%	68%	
Depends		12%	23%	
No good cause		8%	4%	
Don't know		4%	4%	
No information				
Self-monitoring				
Good cause				62%
Depends				12%
No good cause				6%
Not very meaningful				14%
Don't know				6%

In general, it can be concluded from Table 8.22 that subjects show very favorable attitudes toward large-scale implementation of the conservation booklet, feedback, and self-monitoring. (5) No differences with respect to distribution of the conservation booklet are observed between experimental conditions. The somewhat less favorable attitudes of monthly feedback subjects compared to biweekly feedback subjects concerning large-scale implementation of feedback is largely due to a greater emphasis on conditional factors by the former.
Table 8.23 presents findings with respect to H_{21}.

Table 8.23: Correlation Between Attitude Toward Large-Scale Implementation by Public Utility Companies of Conservation Booklet, Feedback, and Self-Monitoring and Subjects' Evaluation of Conservation Booklet, Feedback, and Self-Monitoring

	Evaluation of Conservation Booklet	Evaluation of Feedback	Evaluation of Self-Monitoring
Attitude toward large-scale implementation of conservation booklet	.24*		
Attitude toward large-scale implementation of feedback		.22*	
Attitude toward large-scale implementation of self-monitoring			.43*

*$p = \leq .005$

The findings reported in Table 8.23 support H_{21}: the more positive subjects' evaluation of the conservation booklet, feedback and self-monitoring, the more favorable their attitude toward large-scale implementation of the conservation booklet, feedback and self-monitoring, respectively. The strongest evidence is observed for self-monitoring.
A subsequent analysis revealed that the more positive subjects' opinion about continuation of feedback and self-monitoring, the more favorable their attitude toward large-scale implementation of feedback ($R = .39$, $p = .001$) and self-monitoring ($R = .67$, $p = .001$). Also, there is some evidence that energy attitude change is positively related to subjects' attitude toward implementation of the conservation booklet ($R = .27$, $p = .001$), and self-monitoring ($R = .31$, $p = .006$). Finally, it is found that implementation attitudes are neither sta-

tistically significantly correlated with perceived effectiveness of conservation efforts nor with subjects' relative 1980-81 natural gas consumption.

8.11 Demographics

This section will briefly present some findings with respect to demographic correlates (age, education, socioeconomic status) of a number of cognitive variables investigated in this chapter. Results are summarized in Table 8.24.

Table 8.24: Demographic Correlates of Cognitive Aspects of Experimental Interventions

Cognitive Aspects	Age	Education	Socioeconomic Status
Conservation booklet			
Reading intensity	-.14*	.17***	.18***
Consulting frequency	-.03	.08	-.01
Ascribed effectiveness	-.04	.10	.01
Attitude toward large-scale implementation	-.16***	.03	-.01
Evaluation of booklet	-.03	.01	.01
Feedback			
Reading intensity	-.18**	.21***	.17**
Understanding natural gas consumption trends	.05	-.01	-.07
Understanding electricity consumption trends	.10	-.03	-.08
Ascribed effectiveness	-.10	.10	-.01
Household discussions of feedback letters	-.09	.14*	.18**
Opinion about continuation of feedback	-.19**	.09	.03
Attitude toward large-scale implementation	-.23***	-.09	-.03
Evaluation of feedback	-.19**	.05	.01
Self-monitoring			
Preferred recording frequency	.01	.02	-.03
Understanding natural gas consumption trends	-.19	.16	.44***
Understanding electricity consumption trends	-.36**	.15	.38**
Ascribed effectiveness	-.32*	.02	.28*
Opinion about continuation of self-monitoring	-.18	-.03	.17
Attitude toward large-scale implementation of self-monitoring	-.16	-.07	.06
Evaluation of self-monitoring[1])	-.24	.02	.34**

(Table 8.24 continued)

Cognitive Aspects	Age	Education	Socioeconomic Status
Perceived success of natural gas conservation efforts	-.05	.08	.13*
Perceived success of electricity conservation efforts	.05	-.04	.01
Price knowledge natural gas	-.08	.05	.14***
Price knowledge electricity	-.04	.02	.15***
Energy attitude change	-.12	.11**	.04
Comfort importance	.30***	-.29***	-.15***
Reported loss of comfort	.10*	-.04	.05

Note.

1) composite index of ascribed effectiveness, opinion about continuation of self-monitoring, and attitude toward large-scale implementation of self-monitoring.

* $\underline{p} = \leq .05$
** $\underline{p} = \leq .025$
*** $\underline{p} = \leq .01$

Table 8.24 shows some, though not marked, differences in cognitive aspects of conservation information, feedback and self-monitoring with respect to socio-demographic characteristics. There is a significant tendency for older subjects compared to younger subjects to have read the conservation booklet and feedback letters less intensively, to have a more negative attitude toward large-scale implementation of the booklet as well as of regular energy feedback, to be less convinced of the usefulness of continuation of feedback, and to have less positive evaluations of feedback. Also they express less understanding of consumption patterns as indicated by self-monitoring activities and ascribe less effectiveness to those activities. Finally, older subjects more often emphasize the importance of comfort at home and report some more loss of comfort during last winter than younger subjects. Next, education tends to be statistically significant positively related with reading intensity of conservation booklet and feedback letters, household discussions of feedback letters, and energy attitude change, and negatively with comfort importance. Finally, subjects with higher socioeconomic status tend to have read the booklet and feedback letters more often than subjects with lower socioeconomic status, discussed the feedback letters more often with other household members, report a better understanding of consumption trends as indicated by self-monitoring (though not by feedback), ascribe greater effectiveness to self-monitoring, and have a more positive overall evaluation of self-monitoring. Also socioeconomic status is somewhat positively correlated with energy price knowledge and negatively correlated with comfort importance.

Thus, there is some evidence for sociodemographic differentiation with respect to cognitive aspects of experimental treatments but this evidence is not always consistent between experimental treatments.

8.12 Summary

This chapter further explored the psychological context of the experimental treatments by investigating a number of cognitive aspects related to those treatments. First of all it was concluded that against the background of multiple energy conservation stimuli subjects are exposed to in everyday life, the experimental stimuli cannot be characterized as highly salient. Next, only in the case of the information group was empirical evidence found for the hypotheses (H_{15}, H_{16}) that the more intensive consumers' pre-experimental need for energy conservation information and feedback and the greater their perceived effectiveness of those treatments, the more effective information and feedback are in reducing their energy consumption. Unfamiliarity with feedback and possible interpretation difficulties probably account for this outcome.

It was observed that subjects read the conservation booklet with interest and evaluate this booklet in quite positive terms. Findings show that with respect to natural gas consumption, the greater subjects' interest in the booklet and the more positive their evaluation, the more effective the booklet was in helping them to conserve energy.

Next, it was concluded that although biweekly and monthly energy consumption feedback has been received with interest and is generally positively evaluated, it could not be demonstrated that subjects processed the feedback information in accordance with psychological principles underlying feedback. Results indicate that a significant segment reported difficulties in understanding the feedback information, however, no empirical evidence could be observed for a positive relationship between interest in feedback, comprehensibility of feedback information and conservation trends. Evaluation of feedback turned out to be positively correlated with natural gas conservation. Finally, it was concluded that feedback did not supply subjects with confirmatory evidence about their behavioral performance.

About half of the subjects assigned to the self-monitoring condition complied with the self-monitoring request. High recording frequencies were especially applied by subjects living in insulated dwellings. One could assume in view of this finding that consumers who have invested in insulation or who pay a rent increase for insulation are more interested in observing trends in their household energy usage. Some positive indications were found of the effectiveness of frequent self-monitoring of household energy consumption. Those indications concern observed conservation trends and confirmation of conservation trends.

Next, two hypotheses about the role of energy knowledge were tested. Both hypotheses had to be rejected. First, it appeared that effectiveness of experimental interventions was not highest among subjects with higher (pre-experimental) energy knowledge levels (as hypothesized in H_{17}) but among subjects with lower energy knowledge levels. Secondly, it was found that being exposed to experimental treatments did not increase subjects' knowledge of residential energy matters (as hypothesized in H_{19}).

Also, the hypothesis (H_{18}) about energy attitude change as a consequence of experimental intervention was not supported as the majority of both experimental and control group subjects showed more positive attitudes toward residential energy conservation. Although the findings are not very conclusive, the hypothesis (H_{20}) stating that exposure to experimental treatments does not result in reported loss of comfort, had to be rejected.

Further, some findings were presented which support H_{21} stating that the more positive subjects' evaluation of energy conservation information, feedback, or self-monitoring, the more positive their attitude toward large-scale implementation of those conservation strategies by public utility companies.

Finally, some evidence was provided for sociodemographic differentiation with respect to cognitive aspects of experimental treatments investigated in this chapter, but this evidence is not always consistent between experimental treatments.

Notes

1. Nevertheless, one has to realize that subjects' appraisal of the goal of the post-experimental interview may still result in overestimation of their evaluation of the experimental treatments.

2. See note 2, Chapter 4.

3. Note that subject is not necessarily the same as energy usage recorder.

4. Not surprising, it is found that energy conservation attitude change is positively correlated ($R = .36$; $p = .001$) with perceived success of conservation efforts. However, attitude change is only weakly ($R = .15$, $p = .009$) related to the intention to continue conservation efforts. In turn, perceived success of conservation efforts is not significantly correlated with the intention to continue conservation efforts.

5. Social desirable answers could occur with those questions.

PART IV CONCLUSIONS

9. CONCLUSIONS, POLICY IMPLICATIONS, AND RESEARCH RECOMMENDATIONS

This chapter outlines the main conclusions from this study and formulates some energy conservation policy implications as well as discusses some research recommendations.

At the present time Dutch consumers are provided with feedback only once a year about their household energy consumption (amount of natural gas and electricity consumed in preceding billing year). It can be deduced from behavior change theories that from a psychological point of view such a feedback frequency is highly insufficient for stimulating household energy conservation. One of the central themes of this experimental study concerned the question of what happens if consumers are given more frequent feedback about decreasing or increasing trends in their household energy consumption. The underlying psychological assumption is that regular energy consumption feedback will promote and reinforce conservation efforts. This question was also addressed within the context of findings from quite a number of American studies which showed that regular feedback is able to stimulate household energy conservation. However, most of those studies used daily feedback frequencies which may be effective for psychological reasons, but which - given organizational constraints and cost considerations - obviously are not feasible from a policy point of view. As outlined, due to policy considerations this experiment did not use daily feedback frequencies but applied biweekly and monthly intervals. Also, an attempt was made to overcome a number of observed shortcomings of American studies.
The most important outcome of this experimental field study among approximately 400 consumers is that biweekly or monthly feedback is not more effective in stimulating household energy conservation than providing consumers with energy conservation information or prompting them to monitor their household energy consumption. Therefore, the findings do not support introduction of biweekly or monthly feedback in the Netherlands, at least if the results have generalization value. Considerations with respect to both financial impacts and observed effectiveness do not justify large-scale implementation of such a far-reaching energy conservation policy measure.
One may assume that only in case of highly frequent (e.g. weekly or daily) feedback are consumers able to attribute educational significance to feedback by relating consumption data to behavioral performance. Both options clearly reveal strained relations between psychological effectiveness and policy applicability of highly regular feedback. One interesting possibility to diminish these strained relations is to provide feedback with so-called home energy

monitors. Using simple informational technology such monitors continuously and cumulatively inform consumers about observable increasing or decreasing trends in their household energy usage, visually displayed on the monitor. Scarce research findings report quite positive experiences with home energy monitors, though accurate understanding of displayed consumption trends is often somewhat problematic (cf. Becker & Seligman, 1978; McClelland & Cook, 1978, 1979). A psychologically relevant extension of these monitors is the possibility of combining feedback with goal setting. This combination implies that consumers can choose a specific energy conservation percentage which is stored in the monitor's memory so that consumers receive goal-directed feedback with respect to their behavioral performance (cf. Becker, 1978; Seligman, Becker & Darley, 1981). Further experimentation with home energy monitors could yield promising results, provided that the payback period is also taken into account.

Since the beginning of the seventies, or more precisely after the oil crisis of 1973, utility companies have launched several information campaigns prompting household energy conservation. A review of behavioral studies which experimentally investigated the effectiveness of supplying consumers with energy conservation information generally observed little or no effectiveness of such informational strategies. It was argued, however, that an alternative explanation suggests that this finding might be related to poor design of informational strategies in those studies. In this experiment an attempt was made to develop an informational strategy based on simple but essential principles from communications theory. An energy conservation information booklet was designed which contained practical conservation tips specifically tailored to the target group of this study (housewives living in dwellings with individually metered gas-fired central heating systems). The booklet systematically showed financial consequences (i.e. monetary savings) of energy conservation behaviors. Results indicate that supplying consumers with such information yield small but persistent conservation effects with respect to natural gas consumption.
In view of large-scale distribution possibilities of energy conservation information as well as its relatively low-cost nature, the results of this study lead to a reasonably positive evaluation of supplying consumers with (written) energy conservation information. Again, it has to be emphasized that such information must be tailored to specific target groups and must highlight monetary savings. As far as these monetary savings are concerned, the results show some evidence that conservation information is especially effective for consumers with relatively little understanding of such financial consequences.
In addition, it has to be stressed that other behavioral interventions promoting residential energy conservation (e.g. reinforcement, feedback, legislation,

self-monitoring) can hardly be effective without a supporting conservation information strategy: energy conservation will only be possible if consumers have accurate knowledge of adequate conservation practices. Consequently, those interventions will usually be combined with information strategies, which again emphasizes the importance and perhaps even the indispensability of well-designed energy conservation programs.
As far as energy conservation information is concerned, this study investigated the effectiveness of primarily written information. From communications theory and research it is known that subjects with lower education can be less easily reached with written information. It appears that these groups experience difficulties in interpreting, for instance, numerical information. As this is often a crucial part of energy conservation information, it is desirable to experiment with other methods and channels of information transmission. Of particular interest in this respect are personal home energy audits (Olsen & Cluett, 1979), local media like cable television (Winett & Ester, 1983; Winett, Leckliter, Love, Chinn & Stahl, 1983), and community energy conservation programs (Ester, Gaskell, Joerges, Midden, van Raaij & de Vries, 1984; Joerges et al., 1982).

This study provided some indication for the effectiveness of regular self-monitoring by consumers of observable trends in their household energy consumption, again, however, only with respect to natural gas usage. Obvious advantages of self-monitoring include active consumer involvement and participation, large-scale applicability, and attractiveness in terms of cost considerations. It seems therefore recommendable to further explore self-monitoring possibilities and conceivable extensions, especially with regard to active support (e.g. by utility companies), design adjustments and improvements (recording frequency, type of information to be recorded, facilitation of necessary computations), and possible combination with budget agreements and conservation goal setting.
Next, it was found that subjects living in insulated dwellings in particular complied with the self-monitoring request. Not only did they comply more often but also recorded their meter readings more frequently compared to subjects living in non-insulated dwellings. The following explanation was offered: consumers who invested in home insulation are faced with uncertainty whether this investment balances intended conservation effects. Regular self-monitoring may diminish such feelings of uncertainty by testing the effectiveness of their conservation efforts.
Given these results, a practical policy recommendation would be to combine home insulation programs with active self-monitoring by residents of their household energy usage. Insulation information campaigns could play a significant role in

this respect.
In short, these findings support a plea for an active self-monitoring policy for the residential energy sector.

As far as presently circulating self-recording charts (mainly distributed by utility companies) are concerned, it has to be noted that these charts disproportionally focus consumer attention on the relationship between outside temperature and household energy consumption. The danger exists that consumers are thereby too little encouraged to monitor their own energy behavior, so that it remains unclear in which way energy behavior change is related to lower energy consumption. Against this background it is recommended to investigate ways in which behavior-oriented aspects can be made more salient in self-monitoring procedures.

An analysis of experimental effects showed that exposure to experimental treatments resulted in a 3%-reduction of subjects' natural gas consumption in the experimental period and a 4%-reduction in the post-experimental period. Although these reduction percentages are quite modest at the individual level they may nevertheless be considerable if projected at aggregated levels. Next, it has to be realized that the treatments did not prompt the adoption of new energy conservation technologies which may yield larger consumption reductions.
No statistically significant effects were found for electricity consumption both during and after experimental intervention. These results - which were also found in a more or less similar Dutch experimental study by Midden, Weenig, Houwen, Meter, Westerterp, and Zieverink (1982) may be explained by the fact that household electricity consumption, unlike natural gas usage is a summed measure of quite divergent and heterogeneous elements with makes it difficult to demonstrate small effects. Also, a large part of electricity is used "automatically" by household appliances that are little subject to behavior change. Finally, one has to bear in mind that about 80% of Dutch household energy consumption is used for space heating (mainly by natural gas) purposes.

Sophisticated, well-designed conservation information programs should be focussed on prompting effective energy behavior changes. This study indicates that subjects made a number of divergent energy behavior adjustments which predominantly yielded small conservation effects. Reported energy behavior changes included: lower thermostat settings, shorter heating season, and more attentive energy usage like lowering thermostat when leaving the house, closing hall-door more often, lowering thermostat one hour before going to bed and

using less lighting. In accordance with existing empirical evidence (Cunningham & Lopreato, 1977; Olsen, 1981; Ritsema, Midden & van der Heijden, 1982) it can be concluded that consumers are inclined to perform less radical energy behavior adjustments, i.e. adjustments which evoke least reactance effects.
In view of observed unclear relationships between various energy behavior changes consumers can adopt, it is recommended that future research attempts to identify more structural patterns in order to improve our understanding of possible interrelationships of energy behavior adjustments, which, in turn, may serve as a major input for conservation programs.

Next, in line with many other studies, it was found that consumers generally have little knowledge of residential energy matters. Ellis and Gaskell (1978) introduced the concept of consumer "energy illiteracy" to describe this phenomenon. It could not be demonstrated that supplying subjects' with conservation information systematically improves their energy knowledge. Therefore, it is recommended to investigate ways in which knowledge aspects can be integrated in energy conservation information programs in order to increase consumer energy knowledge.
Results showed that although consumers generally have quite positive attitudes toward household energy conservation, these attitudes are not very deeply rooted. In order to make conservation attitudes less superficial and more salient, conservation information programs could prompt consumers to arrange practical and simple agreements within their household with regard to energy behavior and energy conservation. An advantage of such agreements is that energy usage becomes a discussion subject on the household agenda which may activate social control mechanisms.

Findings indicated that behavioral interventions tested in this study were most effective for consumers with least positive attitudes toward energy conservation and low energy knowledge. The explanation offered suggests that this outcome is related to the fact that the interventions strongly emphasized financial consequences (e.g. monetary savings) of household energy usage and energy conservation. Probably, well-informed consumers with positive conservation attitudes no longer need such information.

Next, it was concluded that the attempt made in this study to explain subjects' intention to conserve energy and reported energy conservation behaviors by their beliefs about consequences of energy conservation and their evaluation of those consequences, as by their beliefs about normative expectations, has been only moderately successful. Apparently, energy conservation intentions and behaviors are complexly determined and difficult to predict. A number of research suggestions was offered to improve this prediction: (a) improved measurement of salient energy conservation beliefs, (b) more precise inclusion of situational variables, (c) unfolding of conservation intentions within separate behavioral contexts, and (d) inclusion of subjects' past behavioral experience with energy conservation.

Quite unexpectedly, it had to be concluded that exposure to experimental treatments did somewhat result in reported loss of comfort. Although comfort preference is an important determinant of consumers' attitude toward household energy conservation (Seligman, Darley & Becker, 1978), similar experiments did not observe loss of comfort (Winett, Hatcher, Fort, Leckliter, Love, Riley & Fishback, 1982; Winett, Leckliter, Love, Chinn & Stahl, 1983). For a correct understanding of our findings it has to be realized that comfort issues were not a substantial part of this study and were only briefly addressed. Nevertheless, comfort matters are crucial issues for both energy and consumer policy. It is therefore recommended to separately investigate a number of energy conservation behaviors which differ in energy consequences but are equal with respect to comfort impacts, and vice versa. Such research may result in practical policy recommendations with regard to possible substitution strategies (Winett, Geller & Everett, 1982).

Finally, a conclusion of vital importance is that interpretation of outcomes from this study would have been quite different when solely concentrating on the survey results. However, the reverse also holds: exclusively examining energy consumption data would have generated little insight into ways in which subjects cognitively processed information supplied by experimental treatments. Combining those two observation methods has been crucial in understanding and evaluating the experimental findings.

Given the correctness of this conclusion it can be recommended to not only evaluate energy conservation programs by survey results but also in combination with examining energy consumption data wherever possible. Omitting this last method could easily yield a systematic bias (more conservation acts reported than actually took place) in evaluating those programs.

10. SUMMARY

The primary goal of this field-experimental study was to investigate the effectiveness of behavioral interventions aimed at promoting residential energy conservation by consumers. The following interventions have been tested: providing consumers with written household energy conservation information, supplying consumers with biweekly or monthly feedback about decreasing or increasing trends in their household energy consumption, and prompting consumers to regularly monitor their energy usage. The effectiveness of those four energy behavior change methods was tested in a field experiment in five towns in the Netherlands among approximately 400 housewives.
This study contains four parts. Part I (Chapters 1 thru 4) discusses the research questions underlying this study as well as their theoretical relevance, gives a detailed review of the existing literature, and formulates the research hypotheses. Part II (Chapter 5) describes the experimental design and the implementation of the experimental conditions. Part III (Chapters 6 thru 8) presents the research findings in view of the research hypotheses. Part IV (Chapters 9 thru 11), finally, summarizes the findings of this study and formulates both policy and research implications and recommendations.

Chapter 1 contains a general introduction to the main theme of this study. It is argued that energy conservation has many advantages over other energy policy options, especially in view of sociopolitical, economic, technical, safety, and environmental considerations. Given the fact that energy behavior is an important determinant of household energy consumption, the thesis is elaborated of the indispensability of a social-behavioral approach of consumer energy conservation. Promoting energy conservation is not simply and solely a technological question but as much a social scientific question. Finally, a more precise formulation is offered of the research question guiding this study which is directed at examining the theoretical and empirical effectiveness of behavior change methods promoting household energy conservation.

Chapter 2 discusses a number of general social scientific theories which may provide a basic insight into conditions in which consumers are willing to conserve energy and factors which impede this willingness. The theories discussed particularly focus on strained relations between short-term individual and long-term collective costs and benefits of individual decisions. Attention is paid to theories about "social traps" (Platt, 1973), "the tragedy of the commons" (Hardin, 1968), prisoner's dilemmas, as well as to Olson's (1965) collective action theory. Next, the literature is reviewed with respect to a number of psychological and sociological determinants or correlates of consumer energy conservation behavior, including specific energy conservation attitudes, general attitudes toward energy scarcity, value orientations, energy knowledge, consumer lifestyle, and sociodemographic characteristics.

Chapter 3 analyzes a number of antecedent and consequence strategies which can be applied to promote consumer energy conservation behavior and examines the validity of underlying behavior change assumptions. The following strategies are discussed: information/education, prompting, modeling, feedback, self-monitoring, positive and negative reinforcement. Next, a systematic and extensive review is presented of behavioral experimental research (mainly American) in this area. This review of the existing literature gives rise to a number of critical notes with respect to both theoretical, methodological, and policy aspects of this developing research tradition.

Chapter 4 legitimizes the choice of behavior change strategies which have been field-experimentally tested in this study (conservation information, biweekly and monthly energy consumption feedback, and self-monitoring of energy meter readings) and clarifies how an attempt has been made to overcome a number of shortcomings detected in existing behavioral energy conservation experiments. Next, the research hypotheses are formulated which guided this study. These hypotheses concern, among other things, the absolute and relative effectiveness of the four selected behavioral interventions, attitude-behavior relationships, energy knowledge, comfort experiences, and segmentation aspects. In total, some twenty hypotheses were tested.

Next, the experimental design is outlined in some detail in Chapter 5. The field experiment was conducted in five towns in the Netherlands. Subjects were approximately 400 housewives recruited from five neighborhoods with physically identical, mainly single-family dwellings supplied with individually metered gas-fired central heating systems. Participants were randomly assigned to either the experimental groups or the control group and were extensively interviewed before and after the experiment. The following behavior change methods were tested:

group 1: received written household energy conservation information only

group 2: received written conservation information plus biweekly household energy consumption feedback group 3: received written conservation information plus monthly household energy consumption feedback

groep 4: received written conservation information plus a request to record their household energy usage (self-monitoring condition)

group 5: received no experimental stimuli and is used as a no-treatment control group

Written conservation information: based on existing information sources a manual on household energy conservation practices was compiled in close cooperation with the Dutch Foundation for Energy Conservation Information (SVEN) and the Dutch Department of Economic Affairs. The manual prompts practical energy behavior changes. The main theme concerns financial benefits of household energy conservation. Both design and content of the manual incorporate basic elements from communications theory and social marketing.
Feedback: every two weeks or every four weeks subjects were provided with information about the amount of natural gas and electricity they consumed in the preceding two or four weeks. In addition, they were informed whether a relative decreasing or increasing trend could be observed in their household energy consumption, compared to usage levels in the preceding year. Financial consequences of those trends were highlighted. Outside temperature corrections were used for natural gas consumption and seasonal corrections for electricity consumption.
Self-monitoring: subjects in this experimental condition received a meter reading recording form together with the request to frequently record their energy usage.
The experimental period lasted from January 1981 till July 1981. In total the biweekly feedback group received feedback 12 times, the monthly feedback group 6 times. During the experimental period subjects in the self-monitoring condition received a recording form three times; each form was sufficient for six weeks.
Next, the chapter provides detailed information about selection of research locations, sampling procedures, recruitment of subjects, design and implementation of experimental conditions, response to pre- and post-experimental (face-to-face) interviews with subjects, and measurement of key-concepts. Finally, a graphic representation of the main research variables is included.

Before presenting the experimental results, one important remark has to be made. A crucial dimension of this study concerns policy applicability of expe-

mentally tested energy behavior change methods. Consequently, a major starting-point involved realistic design of those behavior change methods, especially with respect to policy relevance and policy feasibility. Passive contacts were therefore maintained with subjects and subjects were not rewarded for their participation.
Particularly in case of both feedback conditions, this basic starting point would presumably have direct consequences for the psychological effectiveness of the experimental conditions. For instance, it is known from behavior change research that the more frequent performance feedback is given, the more effective feedback is in stimulating behavior change. For this reason, many American behavioral energy conservation experiments used daily energy consumption feedback frequencies. However, in view of financial costs and organizational constraints the policy relevance of such frequencies is highly disputable. Against the background of these considerations this experiment explicitly opted for much lower feedback frequencies. One obvious disadvantage of this choice is, of course, that almost by definition no spectacular results can be expected. On the other hand, however, a more realistic evaluation may be obtained of the effects of feedback. Nevertheless a strained relationship exists between psychological effectiveness and policy feasibility of behavioral interventions promoting consumer energy conservation. Therefore, the experimental results have to be explicitly interpreted in view of this strained relationship.
Finally, it must be emphasized that this study did not prompt the adoption of new energy conservation technologies but was directed at simple energy conservation behavior changes.

In Chapter 6 the findings from the pre-experimental survey among subjects are discussed. Some hypotheses are tested regarding belief structures with respect to household energy conservation. It is concluded that the attempt made in this study to explain subjects' intention to conserve energy by their beliefs about consequences of energy conservation and their evaluation of those consequences, as well as beliefs about normative expectations, has only been moderately successful. Together these specific energy attitudes explain 30% of the variance in the intention to conserve energy. Apparently, energy conservation intentions are complexly determined and difficult to predict. Four suggestions are offered to improve this prediction: (a) better measurement of salient energy conservation beliefs, (b) more precise inclusion of situational variables, (c) unfolding of conservation intentions within separate behavioral contexts, and (d) inclusion of subjects' past behavioral experience with energy conservation. It is found that both the overall attitude toward household energy conservation and the intention to conserve energy are quite positive in our sample.

The hypothesis (H_1) that specific energy conservation attitudes are better predictors of intentions to conserve energy than general energy attitudes is supported. In correspondance with H_6, it is observed that attitudinal factors are stronger determinants of subjects' intention to conserve energy than normative factors. No empirical support, however, is found for the hypothesis (H_7) that the intention to conserve energy is more influenced by subjects' beliefs about and evaluations of personal consequences of energy conservation than by their beliefs and evaluations of possible social consequences. Both factors correlate about equally strong with the intention to conserve energy.
The phenomenon of "consumer energy illiteracy" is to some degree also observed in this study. Although the correlations are generally weak, the hypothesis (H_8) that there is no significant relationship between consumers' knowledge of residential energy matters and their specific and general energy attitudes and their intention to conserve energy nevertheless has to be rejected.
Next, some data are provided about subjects' attitudes related to the experimental conditions, particularly their pre-experimental needs for energy conservation information and regular energy consumption feedback. Correlational evidence suggests that specific and general energy attitudes, intention to conserve energy, energy knowledge, and needs for conservation information and regular feedback are only moderately or weakly related to subjects' sociodemographic characteristics (age, education, socioeconomic status). Finally, some information is supplied with respect to subjects' attitudes toward some more general energy issues.

Chapter 7 presents the main experimental results. First of all, however, the necessary energy consumption data preparations applied in this study are outlined and some information is given about baseline consumption in the five research locations.
It is found that all experimental interventions (conservation information, biweekly feedback, monthly feedback, self-monitoring) resulted in reduced natural gas consumption. No significant effects are found for electricity usage. An explanation is offered for this finding. The hypotheses (H_9, H_{10}, H_{11}, H_{12}) concerning the assumed effectiveness of the experimental interventions are therefore partially confirmed. No significant differences in effectiveness between experimental conditions are observed. The hypothesis (H_{13}) stating that self-monitoring is relatively most effective and conservation information is least effective is consequently rejected.
In spite of the fact that the experimental interventions tested in this study have only been moderately successful in promoting consumer energy conservation (reductions of 3% - 5%), the amount of cubic meters natural gas "saved" could

nevertheless be considerable when taken community- or nation-wide, especially with low-cost interventions, like conservation information and self-monitoring.

Next, it is found that only a small fraction of subjects' relative 1980-81 natural gas consumption can be explained by their pre-experimental specific and general energy attitudes. Thus, there is only limited support for H_2 and H_4. Also, little empirical evidence is found for the "breeding ground hypothesis" (H_{14}) stating that the effectiveness of experimental interventions is related to subjects' pre-experimental attitudes toward energy conservation.

Quite interestingly, it is observed that if thermostat setting agreements exist within households, specific energy conservation attitudes are more strongly correlated with subjects' relative 1980-81 natural gas consumption. In accordance with H_5 it is concluded that subjects' pre-experimental behavioral intention to conserve energy is a moderate predictor of their adoption of specific energy conservation behaviors. Next, it is found that subjects' pre-experimental attitudes toward both energy conservation and energy scarcity are hardly or not correlated with their involvement in specific energy conservation behaviors, which leaves little or no support for H_3. A number of theoretical and methodological explanations are offered for this finding.

Finally, it is concluded that sociodemographic correlates of the effectiveness of experimental interventions do not reveal a clear pattern which could be used for segmentation purposes.

Chapter 8 extensively examines the psychological context of the experimental treatments by investigating a number of cognitive aspects related to those treatments.

First of all it is concluded that, in view of multiple energy conservation stimuli subjects are exposed to in everyday life, the experimental stimuli cannot be characterized as highly salient. Next, only in the case of the information group is empirical evidence found for the hypotheses (H_{15}, H_{16}) that the more intensive consumers' pre-experimental need for energy conservation information and feedback and the greater their perceived effectiveness of those treatments, the more effective information and feedback are in reducing their energy consumption. Unacquaintedness with feedback and possible interpretation difficulties probably account for this outcome.

From the post-experimental survey (face-to-face interviews) among subjects it can be concluded that subjects read the conservation booklet with interest and evaluate the booklet in quite positive terms. Findings show that with respect to natural gas consumption, the greater subjects' interest in the booklet and the more positive their evaluation, the more effective the booklet was in help-

ing them to conserve energy.
Next, it is observed that although biweekly and monthly energy consumption feedback has been received with interest and is generally positively evaluated, it could not be demonstrated that subjects processed the feedback information in accordance with psychological principles underlying feedback. Results indicate that a significant segment reported difficulties in understanding the feedback information. However, no empirical evidence could be observed for a positive relationship between interest in feedback, comprehensibility of feedback information and conservation trends. Evaluation of feedback turns out to be positively correlated with natural gas conservation. Finally, it is concluded that feedback did not supply subjects with confirmatory evidence about their behavioral performance.
About half of the subjects assigned to the self-monitoring conditions complied with the self-monitoring request. High recording frequencies were particularly applied by subjects living in insulated dwellings. One could assume in view of this finding that consumers who have invested in insulation or who pay a rent increase for insulation are more interested in observing trends in their household energy usage. Some positive indications are found of the effectiveness of frequent self-monitoring of household energy consumption. These indications concern observed conservation trends and confirmation of conservation trends.
Next, two hypotheses about the role of energy knowledge are tested. Both hypotheses have to be rejected. First, it appears that effectiveness of experimental interventions is not highest among subjects with higher (pre-experimental) energy knowledge levels (as hypothesized in H_{17}) but among subjects with lower energy knowledge levels. Secondly, it is found that being exposed to experimental treatments does not increase subjects' knowledge of residential energy matters (as hypothesized in H_{19}).
Also, the hypothesis (H_{18}) about energy attitude change as a consequence of experimental intervention is not supported as the majority of both experimental and control group subjects show more positive attitudes toward residential energy conservation. Although the findings are not very conclusive, the hypothesis (H_{10}) stating that exposure to experimental treatments does not result in reported loss of comfort has quite unexpectedly to be rejected. However, only little loss of comfort is reported by subjects.
Further, some findings are presented which support H_{21} stating that the more positive subjects' evaluation of energy conservation information, feedback, or self-monitoring, the more positive their attitude toward large-scale implementation of these conservation strategies by public utility companies.
Finally, some evidence is provided for sociodemographic differentiation with respect to cognitive aspects of experimental treatments investigated in this

study, but this evidence is not always consistent between experimental treatments.

Chapter 9, in conclusion, summarizes the main findings and conclusions of this study and formulates a number of policy implications and research recommendations.
One of the major results in this study is that biweekly or monthly feedback is not more effective in stimulating household energy conservation than providing consumers with energy conservation information or prompting them to monitor their household energy consumption. Therefore the findings do not support introduction of biweekly or monthly feedback in the Netherlands, at least if the results can be generalized. Considerations with respect to financial costs and organizational requirements do not justify large-scale implementation of such a far-reaching energy conservation policy measure. In view of results from many American studies it may be concluded that feedback is only effective at high (e.g. daily) frequencies. Apparently, a potential conflict exists between psychological effectiveness and policy feasibility of such a behavioral intervention aimed at promoting consumer energy conservation. It is suggested that this problem may not occur with providing feedback through so-called home energy monitors. Further research is recommended with respect to this relatively simple informational technology, especially in view of possible combinations with energy conservation goal-setting. One might also think of providing feedback through teletext or 2-way cable systems.
Next, it is concluded that given large-scale distribution possibilities of energy conservation information, as well as its relatively low-cost nature, the results of this study lead to a reasonably positive evaluation of supplying consumers with (written) conservation information, provided, however, that the information is based on elementary principles from communications theory and social marketing. Also, this study showed some indications of the effectiveness of regular self-monitoring by consumers of observable trends in their household natural gas usage. Obvious advantages of self-monitoring concern active consumer involvement and participation, large-scale applicability, and attractiveness in terms of cost considerations. It is recommended to combine home insulation programs with active self-monitoring by residents. In general, an active self-monitoring policy is proposed for the residential energy sector.
Further research is suggested on alternative methods and channels for the diffusion of energy conservation information. Examples are personal home energy audits, local media like cable television, and integrated community conservation programs.
Following, a number of research recommendations is proposed with respect to

specific issues, including the development of educational programs directed at promoting accurate consumer energy conservation knowledge, examination of possible structural patterns in energy behaviors, as well as further study of potential conflicts between energy conservation and comfort.

Finally, it is indicated that interpretation of outcomes from this study would have been quite different when solely concentrating on either the survey findings or the consumption data. Restricting the analysis to one of those sources would have yielded an interpretation bias.

REFERENCES

Abma, E., Jägers, H.P.M., & van Kempen, G.J. Kernenergie als maatschappelijke splijtstof, een analyse van een protestbeweging. In: P. Ester & F.L. Leeuw (Eds.) Energie als maatschappelijk probleem. Assen: Van Gorcum, 1981.

Acheson, J.M. The lobster fiefs: economic and ecological effects of territoriality in the Maine lobster industry. Human Ecology, 1975, 3. 183-207.

Ajzen, I, & Fishbein, M. Understanding attitudes and predicting social behavior. Englewood Cliffs, N.J.: Prentice-Hall, 1980.

Alchian, A.A., & Demsetz, H. The property rights paradigm. Journal of Economic History, 1973, 33, 16-27.

Altman, I. The environment and social behavior. Monterey, CA: Brooks-Cole, 1975.

Altman, I., & Wohlwill, J. (Eds.). Human behavior and environment: Advances in theory and research. Vol.1, New York: Plenum, 1976.

Altman, I., & Wohlwill, J. (Eds.). Human behavior and environment: Advances in theory and research. Vol. II, New York: Plenum, 1977.

Amstel, R. van., Ester, P., Schijndel, Chr. van., & Schreurs, L.J.M. Energie en huishouden: Verslag van groepsdiscussies onder huisvrouwen. Vrije Universiteit, Instituut voor Milieuvraagstukken, Amsterdam, 1980.

Anderson, C.D., & McDougall, G.H.G. Consumer energy research: An annotated bibliography. Behavioral Energy Research Group, University of British Columbia, Vancouver, 1980.

Anderson, N.H., & Shanteau, J. Weak inference with linear models. Psychological Bulletin, 1977, 84, 1155-1170.

Anderson, R.W., & Lipsey, M.W. Energy conservation and attitudes toward technology. Public Opinion Quarterly, 1978, 42, 17-30.

Annett, J. Feedback and human behaviour. London: Penguin, 1969.

Backer, P., Bruchem, D. van., Hamer, C., Mellink, B., Meijer, G., Moezel, B. van de., Plugge, P., & Westra, H. Vergelijkend vooronderzoek naar gasverbruik in woningen. Eindrapport van de colloquiumgroep GM-I. Delft, Technische Hogeschool, Afdeling Electrotechniek, 1980.

Ban, A.W. van den. Inleiding tot de voorlichtingskunde. Meppel: Boom, 1982.

Bandura, A. Principles of behavior modification. Holt, Rinehart and Winston, Inc., New York, 1969.

Barnaby, D.J., & Reizenstein, R.C. Perspectives on the energy crisis: gasoline prices and the southeastern consumer. Survey of Business, 1975, 2, 28-31.

Bartell, T. Political orientation and public response to the energy crisis. Social Science Quarterly, 1976, 57, 430-436.

Barth, D. Energy demand management in the International Energy Agency (IEA) - results, problems and policies. Paper presented at the Conference on Societal Responses to the Energy Crisis, Dubrovnik, September 1981.

Baum, A., & Singer, J.E. (Eds.) Advances in environmental psychology, Vol. 3, Energy: Psychological perspectives. Hillsdale, N.J.: Lawrence Erlbaum Associates, 1981.

Becker, L.J. Joint effect of feedback and goal setting on performance: A field study of residential energy conservation. Journal of Applied Psychology, 1978, 63, 428-433.

Becker, L.J., & Seligman, C. Reducing air conditioning waste by signalling it is cool outside. Personality and Social Psychology Bulletin, 1978, 4, 412-415.

Bell, P.A., Fisher, J.D., & Loomis, R.J. Environmental psychology. Philadelphia, PA: W.B. Saunders Company, 1978.

Bem, D.J. Self-perception: An alternative interpretation of cognitive dissonance phenomena. Psychological Review, 1967, 74. 188-200.

Bem, D.J. Beliefs, attitudes, and human affairs. Brooks/Cole Publishing Company, Belmont, California, 1970.

Bem, D.J. Self-perception theory. In: L. Berkowitz (Ed.) Advances in experimental social psychology, Vol. 6, New York, Academic Press, 1972.

Bennett, P.D., & Moore, K.D. Consumers' preferences for alternative energy conservation policies: a trade-off analysis. Journal of Consumer Research, 1981, 8, 313-321.

Berkowitz, M.K., & Haines, Jr., G.H. A multi-attribute analysis of consumer attitudes toward alternative space heating modes. Paper presented at the International Conference on Consumer Behavior and Energy Use, Banff, Alberta, September, 1980.

Bittle, R.G., Valesano, R., & Thaler, G. The effects of daily cost feedback on residential energy consumption. Behavior Modification, 1979, 3, 187-202.

Blakely, E. The effect of feedback on residential electrical peaking and hourly kilowatt consumption. Unpublished master's thesis. Department of Psychology, Drake University, Iowa, 1978.

Boer, J. de. Voorlichting over energiebesparing en de notie van het actieve publiek. In: H. Aiking, P. Ester, L. Hordijk & H.E. van de Veen (Eds.) Mozaïek van de Milieuproblematiek, Amsterdam: VU-uitgeverij, 1982.

Boer, J. de., & Ester, P. Consumentengedrag en energiebesparing: Een veldexperimenteel onderzoek naar de effectiviteit van voorlichting, feedback en zelf-registratie. Vrije Universiteit, Instituut voor Milieuvraagstukken, Amsterdam, 1982.

Boer, J. de., Ester, P., Mindell, C., & Schopman, M. Energiebesparingsprogramma's ten behoeve van consumenten in Nederland. Stichting Wetenschappelijk Onderzoek Konsumentenaangelegenheden, Den Haag/Instituut voor Milieuvraagstukken, Vrije Universiteit, Amsterdam, 1982 a.

Boer, J. de., Ester, P., Mindell, C., & Schopman, M. Consumer energy conservation programs in the Netherlands. Paper presented at the International Conference on Consumer Behavior and Energy Policy, Noordwijkerhout, September 1982b.

Bosma, S., & Kok, G. Studies in attitude en gedrag 6: Isolatiegedrag. Een onderzoek naar woningisolatie. Heymans Bulletin, HB-82-574-EX, Rijksuniversiteit Groningen, 1982.

Bowman, C., & Fishbein, M. Understanding public reaction to energy proposals: An application of the Fishbein model. Journal of Applied Social Psychology, 1978, 8, 319-340.

Brechner, K.C. An experimental analysis of social traps. Journal of Experimental Social Psychology, 1979, 13, 552-564.

Brechner, K.C., & Linder, D.E. A social trap analysis of energy distribution systems. In: A. Baum & J.E. Singer (Eds.). Advances in environmental psychology, Volume 3 Energy: Psychological perspectives. Hillsdale, N.Y.: Lawrence Erlbaum Associates, 1981.

Bruggink, J.J.C. De werkgelegenheidseffecten van het Nationaal Isolatieprogramma. Energie Studie Centrum, Stichting Energieonderzoek Centrum Nederland (ESC-15), Petten, 1981.

Brunner, J.A., & Bennett, G.F. Coping with the energy shortage: perceptions and attitudes of metropolitan consumers. Journal of Environmental Systems, 1977, 6, 253-268.

Buchanan, J.M. The demand and supply of public goods. Chicago: Rand McNally & Company, 1968.

Bultena, G. Public response to the energy crisis. A study of citizens' attitudes and adaptive behaviors. Unpublished manuscript, Iowa State University, Department of Sociology and Anthropology, Sociology Report 130, Ames, Iowa: 1976.

Bupp, I.C. The nuclear stalemate. In: R. Stobaugh & D. Yerkin (Eds.), Energy future: Report of the energy project at the Harvard Business School. New York: Random House, 1979.

Burgess, Ph.M., & Robinson, J.A. Alliances and the theory of collective action: A simulation of coalition processes. Midwest Journal of Political Science, 1969, 194-218.

Carlyle, J.J., & Geller, E.S. Behavioral approaches to reducing residential energy consumption: A critical review. Unpublished manuscript, Virginia Polytechnic Institute and State University, Department of Psychology, Blacksburg: VA, 1979.

Cass, B.C., & Edney, J.J. The Commons Dilemma: A simulation testing the effects of resource visibility and territorial division. Human Ecology, 1978, 6, 371-386.

Catton, W.R., Jr., & Dunlap, R.E. Environmental sociology: A new paradigm. The American Sociologist, 1978, 13, 41-49.

Catton, W.R., Jr., & Dunlap, R.E. A new ecological paradigm for post-exuberant sociology. American Behavioral Scientist, 1980, 24, 15-47.

Chamberlin, J.R. The logic of collective action: Some experimental results. Behavioral Science, 1978, 23, 441-445.

Cialdini, R.B., Petty, R.E., & Cacioppo, J.T. Attitude and attitude change. Annual Review of Psychology, 1981, 32, 357-404.

Cone, J.D., & Hayes, S.C. Environmental problems/Behavioral solutions. Monterey, CA: Brooks/Cole Publishing Company, 1980.

Cook, S.W. Promoting energy conservation in master-metered buildings. Paper presented at the 1978 annual meeting of the Human Factors Society, Detroit, 1978.

Cook, S.W., & Berrenberg, J.L. Approaches to encouraging conservation behavior: A review and conceptual framework. Journal of Social Issues, 1981, 37, 73-107.

Cottrell, F. Energy and society: The relation between energy, social change, and economic development. New York, McGraw-Hill, 1955.

Craig, C.S., & McCann, J.M. Assessing communication effects on energy conservation. Journal of Consumer Research, 1978, 5, 82-88. (a)

Craig, C.S., & McCann, J.M. The impact of persuasive communications on energy conservation. Energy Systems and Policy, 1978, 2, 433-447 (b).

Craik, K.H., & Zube, E.H. (Eds.). Perceiving environmental quality. New York: Plenum, 1976.

Crossly, D.J. Barriers to household energy conservation. Paper presented at the International Conference on Consumer Behavior and Energy Policy, Noordwijkerhout, September 1982.

Cunningham, W.H., & Joseph, B. Energy conservation, price increases and payback periods. In: H.R. Hunt (Ed.) Advances in consumer research, vol. 5. Chicago: Association for Consumer Research, 1978.

Cunningham, W.H., & Lopreato, S.C. Energy use and conservation incentives: A study of the Southwestern United States. New York: Praeger Press, 1977.

Curtin, R. Consumer adaptation to energy shortages. Journal of Energy and Development, 1976, 2, 38-59.

Darley, J.M. Energy conservation techniques as innovations, and their diffusion. Energy and Buildings, 1978 1, 339-343.

Darley, J.M., & Beniger, J.R. Diffusion of energy-conserving innovations. Journal of Social Issues, 1981, 37, 150-171.

Darmstadter, J., Dunkerley, J., & Alterman, J. How industrial societies use energy: A comparative analysis. Baltimore, Maryland: The John Hopkins University Press, 1977.

Davidson, A., & Jaccard, J. Population psychology: A new look at an old problem. Journal of Personality and Social Psychology, 1975, 31, 1073-1082.

Dawes, R.M., & McTavish, J. Effect of communication and assumptions about the other people on pro-social and anti-social behavior in a Commons Dilemma situation. Paper presented at the West Coast Conference on Small Group Research, April 1975.

Dawes, R.M., McTavish, J., & Shaklee, H. Behavior, communication, and assumptions about other people's behavior in a commons dilemma situation. Journal of Personality and Social Psychology, 1977, 35, 1-11.

De Boer, C. The polls: nuclear energy. Public Opinion Quarterly, 1977, 41, 402-411.

Dillman, D.A., Rosa, E.A., & Dillman, J.J. Lifestyle and home energy conservation in the United States: The poor accept lifestyle cutbacks while the wealthy invest in conservation. Journal of Economic Psychology, 1983, 299-315.

Doel, J. van den. Demokratie en welvaartstheorie. Een inleiding in nieuwe politieke ekonomie. Alphen aan den Rijn: Samson Uitgeverij, 1975.

Doel, J. van den. Demokratie en welvaartstheorie. Tweede geheel herziene druk. Alphen aan den Rijn, Samson Uitgeverij, 1978.

Duncan, O.D. Sociologists should reconsider nuclear energy. Social Forces, 1978, 57, 1-22.

Dunlap, R.E. Paradigmatic change in social science. From human exemptionalism to an ecological paradigm. American Behavioral Scientist, 1980, 24, 5-14.

Dunlap, R.E., & Catton, W.R., Jr. Environmental sociology: A framework for analysis. In: T. O'Riordan & R.C. D'Arge (Eds.), Progress in resource management and environmental planning, Vol. I, Chicester: John Wiley & Sons, 1979 (a).

Dunlap, R.E., & Catton, W.R., Jr. Environmental sociology, Annual Review of Sociology, 1979, 5, 243-273. (b).

Durand, R.M. A study of Alabama consumer attitudes toward the energy crisis. Unpublished manuscript, The University of Alabama, School of Mines and Energy Development, 1979.

Durkheim, E. Les règles de la méthode sociologique. Paris: Presses Universitaires de France, 1895 (16th edition 1967).

Eagly, A.H., & Himmelfarb, S. Attitudes and opinions. Annual Review of Psychology, 1978, 29, 517-554.

Edney, J.J. The Commons Problem, Alternative perspectives. American Psychologist, 1980, 35, 131-150.

Edney, J.J. Paradoxes on the commons: Scarcity and the problem of equality. Journal of Community Psychology, 1981, 9, 3-34.

Edney, J.J., & Harper, C.S. The Commons Dilemma: A review of contributions from psychology. Environmental Management, 1978, 2, 491-507. (a)

Edney, J.J., & Harper, C.S. Heroism in a resource crisis: A simulation study. Environmental Management, 1978, 2, 523-527. (b)

Edney, J.J., & Harper, C.S. The effects of information in a resource management problem: A social trap analog. Human Ecology, 1978, 6, 387-395. (c)

Eiser, J.R., & Pligt, J. van der. Beliefs and values in the nuclear debate. Journal of Applied Social Psychology, 1979, 9, 524-536.

Elgin, D. Voluntary simplicity. Toward a way of life that is outwardly simple, inwardly rich. New York: William Morrow and Co., Inc., 1981.

Elgin, D., & Mitchell, A. Voluntary simplicity. The Co-Evolution Quarterly, 1977, 5-18.

Ellis, P., & Gaskell, G. A review of social research on the individual energy consumer. Unpublished manuscript, London School of Economics and Political Science, Department of Social Psychology, London, 1978.

Ester, P. Attitudes of the Dutch population toward alternative life styles and environmental deterioration. Paper presented at the Conference on Alternative Ways of Life, United Nations University, Cartigny, April 1978.

Ester, P. Methoden ter bevordering van milieuvriendelijk en energiebewust consumptief gedrag. Vrije Universiteit, Instituut voor Milieuvraagstukken, Amsterdam, Werknota 108, 1979. (a)

Ester, P. Milieubesef en milieugedrag: een sociologisch onderzoek naar attitudes en gedragingen van de Nederlandse bevolking met betrekking tot het milieuvraagstuk. Vrije Universiteit, Instituut voor Milieuvraagstukken, Amsterdam, 1979. (b)

Ester, P. (Ed.). Sociale aspecten van het milieuvraagstuk. Assen: Van Gorcum, 1979 (c)

Ester, P. Environmental concern in the Netherlands. In: T. O'Riordan & R.K. Turner (Eds.), Progress in resource management and environmental planning, Vol. 3, Chichester, 1981.

Ester, P., & Boer, J. de. Energiebesparing in huishoudens: Verslag van een vooronderzoek. Vrije Universiteit, Instituut voor Milieuvraagstukken, Amsterdam, 1980.

Ester, P., Gaskell, G., Joerges, B., Midden, C.J.H., Raaij, W.F. van., & Vries, Th. de. (Eds.). Consumer behavior and energy policy. Amsterdam: North-Holland, 1984.

Ester, P., & Leeuw, F.L. Theoretische sociologie en maatschappelijke vraagstukken, individueel handelen en participatie in de voorziening van collectieve goederen. Mens en Maatschappij, 1978, 53, 5-29.

Ester, P., & Leeuw F.L. (Eds.). Energie als maatschappelijk probleem. Assen: Van Gorcum, 1981.

Ester, P., & Leroy, P. Sociologie en het Milieuvraagstuk: Agendapunten voor sociaal-wetenschappelijk milieu-onderzoek. Paper presented at the Sociologendagen 1984, Free University, Amsterdam, April, 1984.

Ester, P., Linden J.W. van der., & Pligt, J. van der. Meningsvorming rond kernenergie. Intermediair, 1982, 17.

Ester, P., & Meer, F. van der. Sociaal-psychologische determinanten van individueel milieugedrag: schets van een gedragsmodel. In: P. Ester (Ed.) Sociale aspecten van het milieuvraagstuk, Assen: Van Gorcum, 1979. (a)

Ester, P., & Meer, F. van der. Milieubesef en milieugedrag: enkele onderzoeksbevindingen. In: P. Ester (Ed.) Sociale aspecten van het milieuvraagstuk, Assen: Van Gorcum, 1979. (b)

Ester, P., & Meer, F. van der. Determinants of individual environmental behavior: An outline of a behavioral model and some research findings. The Netherlands' Journal of Sociology, 1981, 18, 57-94.

Ester, P., Rooij, G.M. de., & Schreurs, L.J.M. Nuclear power, public opinion, and political party preferences in the Netherlands. Unpublished manuscript, Free University, Institute for Environmental Studies.

Ester, P., & Winett, R.A. Toward more effective antecedent strategies for environmental programs. Journal of Environmental Systems, 1982, 11, 201-222.

Fairweather, G.W., & Tornatzky, L.G. Experimental methods for social policy research. Oxford: Pergamon Press, 1977.

Farhar, B., Weir, P., Unseld, C., & Burns, B. Public opinion about energy: a literature review. Golden, Colorado: Solar Energy Research Institute, 1979.

Farhar-Pilgrim, B., & Shoemaker, F.F. Campaigns to affect energy behavior. In: R.E. Rice & W.J. Paisley (Eds.), Public communication campaigns. Beverly Hills: Sage Publications, 1981

Fazio, R.H., Chen J.M., McDonel, E.C., & Sherman, S.T. Attitude accessibility, attitude-behavior consistency, and the strength of the object-evaluation association. Journal of Experimental Social Psychology, 1982, 18, 339-357.

Fazio, R.H., & Zanna, M.P. Attitudinal qualities relating to the strength of the attitude-behavior relationship. Journal of Experimental Social Psychology, 1978, 14, 398-408. (a)

Fazio, R.H., & Zanna, M.P. On the predictive validity of attitudes: The role of direct experience and confidence. Journal of Personality, 1978, 46, 228-243. (b)

Fazio, R.H., & Zanna, M.P. Direct experience and attitude-behavior consistency. In: L. Berkowitz (Ed.), Advances in experimental social psychology (Vol. 14). New York: Academic Press, 1981.

Fishbein, M., & Ajzen, I. Beliefs, attitude, intention and behavior: An introduction to theory and research. Reading, Mass.: Addison-Wesley, 1975.

Fishbein, M., & Coombs, F.S. Basis for decision: An attitudinal analysis of voting behavior. Journal of Applied Social Psychology, 1974, 4, 95-124.

Foddy, W.H. Shared resources and ecological catastrophes. The Australian and New Zealand Journal of Sociology, 1974, 10, 154-163.

Foley, G. The energy question. Harmondsworth, Middlesex: Penguin Books, 1976.

Frankena, F. Public reception of large-scale renewable energy technologies: A case study of the Hersey, Michigan wood-electric power plant controversy. Paper presented at the International Conference on Consumer Behavior and Energy Policy, Noordwijkerhout, September 1982.

Frohlich, N., & Oppenheimer, J.A. Modern political economy. Englewood Cliffs, N.J.: Prentice-Hall, Inc., 1978.

Frohlich, N., Oppenheimer, J.A., & Young, O.R. Political leadership and collective goods. Princeton, N.J.: Princeton University Press, 1971.

Furuboth, E.G., & Pejovich, S. (Eds). The economics of property rights. Cambridge, Mass.: Ballinger, 1974.

Gaskell, G., & Ellis, P. Energy conservation: A psychological perspective on a multidisciplinary phenomenon. In: P. Stringer (Ed.), Confronting social issues: Applications of social psychology, Vol. 1, European monographs in social psychology 28, London: Academic Press, 1982.

Gaskell, G., Ellis, P., & Pike, R. The energy literate consumer: The effects of consumption feedback and information on beliefs, knowledge and behaviour. Unpublished manuscript, Department of Social Psychology, The London School of Economics and Political Science, London, 1980.

Geller, E.S. Applications of behavioral analysis for litter control. In: D. Glenwick & L. Jason (Eds.) Behavioral community psychology: Progress and prospects. New York: Praeger, 1980.

Geller, E.S., Winett, R.A., & Everett, P.B. Preserving the environment: New strategies for behavior change. Elmsford, NY: Pergamon Press, Inc., 1982.

Geradts, F., & Geradts, J. Energie, ... kennis en gedrag. Een inventarisatie van ons energiebewustzijn. Intermediair, 1978, 14, 39-43.

Gerlach, L.P., Renz, D.O., & Brown, J.B. Key leverage factors in the technology delivery system for solar heating and cooling. Unpublished manuscript, Minneapolis, M.I.: University of Minnesota, Department of Anthropology, 1979.

Gottlieb, D. Texan's response to President Carter's energy proposals. In: S. Warkov (Ed.) Energy policy in the United States: Social and behavioral dimensions. New York: Praeger Press, 1978.

Gottlieb, D., & Matre, M. Sociological dimensions of the energy crisis: A follow-up study. Unpublished manuscript, University of Houston, Energy Institute, Houston: Texas, 1976.

Griffin, J.M. Energy conservation in OECD: 1980-2000. Cambridge, MA: Ballinger Publishing Company, 1979.

Groenewegen, G.G. Progressieve tarieven. Gas, 1980, 100, 103-109.

Grzelak, J., & Tyszka, T. Some preliminary experiments on cooperation in N-person games. Polish Psychological Bulletin, 1974, 5, 81-91.

Hanna, S. Evaluation of energy saving investments. Journal of Consumer Affairs, 1978, 12, 63-75.

Hardin, G. The tragedy of the commons. Science, 1968, 162, 1243-1248.

Hardin, G., & Baden, J. (Eds.). Managing the commons. San Francisco, CA: W.H. Freeman and Company, 1977.

Harris, C.L. The remorseless working of things: A Commons tragedy in environmental resource management. In: A.Seidel & S. Danford (Eds.) Environmental design: Research, theory & application, Proceedings of E.D.R.A., 1979, 10, Buffalo, N.Y.

Hayes, D. Energy: The case for conservation. Washington: Worldwatch Institute, Paper no. 4, 1976.

Hayes, S.C., & Cone,. J.D. Reducing residential electrical energy use: Payments, information, and feedback. Journal of Applied Behavior Analysis, 1977, 10, 425-435.

Hayes, S.C., & Cone, J.D. Reduction of residential consumption of electricity through simple monthly feedback. Journal of Applied Behavior Analysis, 1981, 14, 81-88.

Hayes, S.C. The effects of monthly feedback, rebate billing, and consumer directed feedback on the residential consumption of electricity. Unpublished doctoral dissertation, West Virginia University, 1977.

Head, J.G. Public goods and public policy. Public Finance, 1962, 17, 197-219.

Heberlein, T.A. The three fixes: Technological, cognitive, and structural. In: D.R. Field, J.C. Barron, & B.F. Long (Eds.), Water and community development: Social and economic perspectives. Ann Arbor, MI: Ann Arbor Science Publishers, Inc., 1974.

Heberlein, T.A. Conservation information: The energy crisis and electricity consumption in an apartment complex. Energy Systems and Policy, 1975, 1, 105-117.

Heberlein, T.A., & Warriner, G.K. The influence of price and attitude on shifting residential electricity consumption from on to off-peak periods. Paper presented at the International Conference on Consumer Behavior and Energy Policy, Noordwijkerhout, September 1982.

Heertje, A. Echte economie. Misverstanden over en misstanden in de economie. Amsterdam: De Arbeiderspers, 1977.

Heimstra, N.W., & McFarling, L.H. Environmental psychology. Monterey, CA.: Brooks-Cole, 1974.

Hemrica, E. Inkomen en gasverbruik van gezinshuishoudingen. Gas, 1981, 101, 67-74.

Heslop, L.A., Moran, L., & Cousineau, A. "Consciousness" in energy conservation behavior: An exploratory study. Journal of Consumer Research, 1981, 8, 299-305.

Hill, R.J. Attitudes and behavior. Unpublished manuscript, The University of Oregon, Institute for Social Science Research, Working Paper No. 36, Eugene: Oregon, 1980.

Hoogerwerf, A. Bèta of gamma? Wetenschapsbeleid discrimineert maatschappijwetenschappen. Intermediair, 1980, 16, no.5, 1-9.

Ilgen, D.R., Fisher, C.D., & Taylor, M.S. Consequences of individual feedback on behavior in organizations. Journal of Applied Psychology, 1979, 64, 349-371.

Illich, I. Energy and equity. New York, Harper and Row, 1973.

Institute for Energy Analysis, Oak Ridge Associated Universities. U.S. energy and economic growth, 1975-2000. Oak Ridge, 1976.

International Institute for Applied System Analysis. Energy in a finite world: Paths to a sustainable future. By J. Anderer, A. McDonald & N. Nakicenovic. Project leader: W. Häfele. Cambridge, MA.: Ballinger Publishing Company,1981.

Jerdee, T.H., & Rosen, B. Effects of opportunity to communicate and visibility of individual decisions on behavior in the common interest. Journal of Applied Psychology, 1974, 59, 712-716.

Joerges, B. Consumer energy research: An international bibliography. International Institute for Environment and Society, Science Center Berlin, 1979.

Joerges, B. a.o. Consumer Energy Conservation Policies: A multi-national study. Fourth interim report. International Institute for Environment and Society, Science Center Berlin, 1982.

Joerges, B., & Olsen, M.E. Policies for promoting energy conservation: An American European perspective. International Institute for Environment and Society, Science Center Berlin, 1979.

Johnston, W.J., Cooper, M.C., & Page, Jr., T.J. Consumer energy usage and the tragedy of the commons phenomenon. Columbus, OH.: The Ohio State University, Faculty of Marketing, College of Business Administration, 1981.

Kasperson, R.E;, Berk, G., Pijawka, D., Sharaf, A.B., & Wood, J. Public opposition to nuclear energy: Retrospect and prospect. Science, Technology & Human Values, 1980, 5, 11-23.

Katzev, R., Cooper, L., & Fisher, P. The effect of feedback and social reinforcement on residential electricity consumption. Journal of Environmental Systems, 1981, 10, 215-227.

Kazdin, A.E. Self-monitoring and behavior change. In: M.J. Mahoney & C.E. Thoresen (Eds.), Self-control: Power to the person. Monterey, Ca: Brooks/Cole Publishing Company, 1974.

Kelley, H.H., & Grzelak, J. Conflict between individual and common interest in a n-person relationship. Journal of Personality and Social Psychology, 1972, 21, 190-197.

Kilkeary, K. The energy crisis and decision-making in the family. National Technical Information Service, PB-238 783, 1975.

Klein, H.J. Sonnenenergie - Nutzung aus Anwendersicht. Unpublished manuscript, Karlsruhe: University of Karlsruhe, Department of Sociology, 1979.

Kohlenberg, R., Philips, T., & Proctor, W. A behavioral analysis of peaking in residential electrical-energy consumers. Journal of Applied Behavior Analysis, 1976, 9, 13-18.

Kok, G.J. Attitudes en energiebewust gedrag. In: P. Ester & F.L. Leeuw (Eds.) Energie als maatschappelijk probleem, Assen: Van Gorcum, 1981.

Kok, G.J., Abrahamse, M, Douma, P., Langejan, A., Sietsma, H., Slob, A., & Vries, H. de. Studies in attitudes en gedrag 2: Attitudes, sociale normen en energiebesparend gedrag. Heymans Bulletins, HB-79-437-EX, Rijksuniversiteit Groningen, 1979.

Krusche, H. Alternative Energietechnologien für private Haushalte - die Diffusion von Sonnenkollektoranlagen. In: B. Joerges (Ed.), Verbraucherverhalten und Umweltbelastung. Materialien zu einer verbraucherorientierten Umweltpolitik. Meisenheim: Anton Hains, 1979.

LaBay, D.G., & Kinnear, T.C. Exploring the consumer decision process in the adoption of solar energy systems. Paper presented at the International Conference on Consumer Behavior and Energy Use, Banff, Alberta, September 1980.

LaSOM (Landelijke Stuurgroep Onderzoek Milieuhygiëne). Signalering en evaluatie van oorzaken en effecten van milieuveranderingen. LaSOM reeks, nr. 5 en 6, 1979.

LaSOM Aanzet voor een programma in hoofdlijnen. Nationaal Programma Onderzoek Milieuhygiëne. LaSOM reeks, nr. 11, 1981.

Leach, G. A low energy strategy for the United Kingdom. The International Institute for Environment and Development, London, 1979.

Leeuw, F.L., & Ester, P. Programmering van sociaal-wetenschappelijk energieonderzoek, toegelicht aan de hand van het vraagstuk van acceptatie van energiebeleid. In: P. Ester en F.L. Leeuw (Eds.) Energie als maatschappelijk probleem. Assen: Van Gorcum, 1981.

Leonard-Barton, D. The role of interpersonal communication networks in the diffusion of energy conserving practices and technologies. Paper presented at the International Conference on Consumer Behavior and Energy Use, Banff, Alberta, September 1980.

Leonard-Barton, D. Voluntary simplicity lifestyles and energy conservation. Journal of Consumer Research, 1981, 8, 243-252.

Leonard-Barton, D., & Rogers, E.M. Adoption of energy conservation among California homeowners. Unpublished manuscript, Stanford University Institute for Communication Research, 1979.

Liebrand, W.B.G. Experimentele psychologie en het energievraagstuk. In: P. Ester en F.L. Leeuw (Eds.) Energie als maatschappelijk probleem, Assen: Van Gorcum, 1981.

Liebrand, W.B.G. Interpersonal differences in social dilemmas; A game theoretical approach. Groningen: Rijskuniversiteit, 1982 (dissertation).

Linden, J.W. van der., Ester, P. en Pligt, J. van der. Kernenergie en publieke opinie: Een onderzoek naar achtergronden van houdingen ten aanzien van kernenergie en de invloed hierop van het wonen bij een kerncentrale. In: H. Aiking, P. Ester, L. Hordijk & H.E. van de Veen (Eds.) Mozaïek van de milieuproblematiek, bundel ter gelegenheid van het 10-jarig bestaan van het Instituut voor Milieuvraagstukken. Amsterdam: Vrije Universiteit, 1981.

Lindenberg, S. De structuur van theorieën van kollektieve verschijnselen. In: W. Arts, S. Lindenberg & R. Wippler (Eds.) Gedrag en structuur. Rotterdam: Universitaire Pers, 1976.

Lipsey, M.W. Personal antecedents and consequences of ecologically responsible behavior: A review. JSAS Catalog of Selected Documents in Psychology, 1977, 7 (MS. 1521).

Lloyd, K.E. Reactions to a forthcoming energy shortage, a topic in behavioral ecology. In: G.L. Martin & J.G. Osborne (Eds.) Helping in the community, behavioral applications. New York: Plenum Press, 1980.

Locke, E.A. The relationship of intentions to level of performance. Journal of Applied Psychology, 1966, 50, 60-66.

Locke, E.A. Relationship of goal level to performance level. Psychological Reports, 1967, 20, 1068.

Locke, E.A. Toward a theory of task motivation and incentives. Organizational Behavior and Human Performance, 1968, 3, 157-189.

Locke, E.A., Saari, L.M., Shaw, K.N., & Latham, G.P. Goal setting and task performance: 1969-1980. Psychological Bulletin, 1981, 90, 125-152.

Lopreato, S.C., & Meriweather, M.W. Energy attitudinal surveys: summary, annotations, research, recommendations. (Final report) Center for Energy Studies, University of Texas, Austin, TX, 1976.

Luce, R.D., & Raiffa, H. Games and decision: Introduction and critical survey. London: John Wiley and sons, 1957.

Lulofs, J.G. Theorie van het collectieve handelen (I). De theorie van Olson. Mens en Maatschappij, 1978, 53, 139-171.

Lulofs, J.G. Theorie van het collectieve handelen (II). Kritische kanttekeningen bij de theorie van Olson. Mens en Maatschappij, 1978, 53, 410-441.

Macey, S.M., & Brown, M.A. Residential energy conservation: The role of past experience in repetitive household behavior. Environment and Behavior, 1983, 15, 123-141.

Maidique, M.A. Solar America. In: R. Stobaugh & D. Yergin (Eds.), Energy future: Report of the Energy Project at the Harvard Business School. New York: Random House, 1979.

Maloney, M.P., & Ward, M.P. Ecology: Let's hear from the people. American Psychologist, 1973, 28, 583-586.

Maloney, M.P., Ward, M.P., & Braught, G.N. A revised scale for the measurement of ecological attitudes and knowledge. American Psychologist, 1975, 30, 787-790.

Marwell, G., & Ames, R.E. Experiments on the provision of public goods. I. Resources, interest, group size, and the free-rider problem. American Journal of Sociology, 1979, 84, 1335-1360.

Marwell, G., & Ames, R.E. Experiments on the provision of public goods. II. Provision points, stakes, experience, and the free-rider problem. American Journal of Sociology, 1980, 85, 926-937.

Mazur, A. Opposition to technological innovation. Minerva, 1975, 13, 58-81.

Mazur, A. & Rosa, E. Energy and life-style. Science, 1974, 186, 607-610.

McClelland, L., & Belsten, L. Promoting energy conservation in university dormitories by physical, policy, and resident behavior changes. Journal of Environmental Sytems, 1979-80, 9, 29-38.

McClelland, L., & Canter, R.J. The resident utility billing system: A method of reducing energy waste in master-metered rental housing. Proceedings of the third National Conference and Exhibition on Technology for Energy Conservation, 1979, 23-25.

McClelland, L., & Canter, R.J. Psychological research on energy conservation: Context, approaches, methods. In: A. Baum & J. Singer (Eds.) Advances in environmental psychology, Vol. 3, Energy: Psychological perspectives. Hillsdale, N.J.: Lawrence Erlbaum Associates, 1981.

McClelland, L., & Cook, S.W. Energy conservation effects of continuous in-home feedback in all-electric homes. University of Colorado, Institute of Behavioral Science, Boulder, 1978.

McClelland. L., & Cook, S.W. Energy Conservation effects of continuous in-home feedback in all-electric homes. Journal of Environmental Systems, 1979, 9, 169-173.

McClelland, L., & Cook, S.W. Energy conservation in university buildings: Encouraging and evaluating reductions in occupants' electricity use. Evaluation Review, 1980, 4, 119-113. (a)

McClelland, L., & Cook, S.W. Promoting energy conservation in master-metered apartments through group financial incentives. Journal of Applied Social Psychology, 1980, 10, 19-31. (b)

McDougall, G.H.G., & Anderson, C.D. Consumer energy research: An annotated bibliography, Volume II. Wilfried Laurier University/University of Manitoba, 1982.

Meer, F. van der. Attitude en milieugedrag. Rijksuniversiteit Leiden, 1981. (dissertation).

Meer, F. van der., & Berghuis, A.C. Attitudes en instrumentaliteit van gedrag. Gedrag, 1976, 4, 24-38.

Melber, B.D., Nealy, S.M., Hammersla, J., & Rankin, W.L. Nuclear power and the public: Analysis of collected survey research. Seattle, WA: Human Affairs Research Centers, 1977.

Meyer, L.A. Energiebespraing in de sociale woningbouw, besparing op ruimteverwarming in theorie en praktijk. Rijksuniversiteit Groningen, 1981. (dissertation).

Meyer-Abich, K.M. (Ed.) Energieeinsparung als neue Energiequelle: Wirtschaftspolitische Möglichkeiten und alternative Technologien. München: Carl Hanser Verlag, 1979.

Midden, C.J.H., Meter, J.E., Weenig, M.E., & Zieverink, H.J.A. Using feedback, reinforcement and information to reduce energy consumption in households: A field experiment. Journal of Economic Psychology, 1983, 65-86.

Midden, C.J.H., & Ritsema, B.S.M. The meaning of normative processes for energy conservation. Journal of Economic Psychology, 1983, 37-55.

Midden, C.J.H., Weenig, W.H., Houwen, R.J., Meter, J.E., Westerterp, G.A., & Zieverink, H.J.A. Energiebesparing door gedragsbeïnvloeding. Den Haag: VUGA uitgeverij, 1982.

Mikulas, W.L. Behavior modification: An overview. New York: Harper & Row, 1972.

Milstein, J.S. Attitudes, knowledge and behavior of American consumers regarding energy conservation with some implications for governmental action. Paper presented at the National Meeting of the Association for Consumer Research, 1976.

Milstein, J.S. How consumers feel about energy: Attitudes and behavior during the winter and spring 1976-1977. In: S. Warkov (Ed.) Energy policy in the United States; Social and behavioral dimensions. New York, Praeger Press, 1978.

Mitchell, T.R., & Biglan, A. Instrumentality theories: Current uses in psychology. Psychological Bulletin, 1971, 76, 432-454.

Moos, R.H., & Insel, P.M. (Eds.) Issues in social ecology: human milieus. Palo Alto, CA.: National Press, 1974.

MOPPS (Market oriented program planning study) Final report, vol. 1. Integrated summary. Washington, D.C.: U.S. Department of Energy, 1977.

Morrison, B.M., Keith, J., & Zuiches, J.G. Impacts on household energy consumption: An empirical study of Michigan families. In: C. Unseld, D. Morrison, D. Sills & C.P. Wolf (Eds.) Sociopolitical effects of energy use and policy. Washington D.C.: National Academy of Sciences, 1979.

Morrison, D.E., Equity impacts of some major energy alternatives. In: S. Warkov (ed.) Energy policy in the United States: Social and behavioral dimensions. New York: Praeger Press, 1978.

Murray, J.R., Minor, M.J., Bradburn, N.M., Cotterman, R.F., Frankel, M., & Pisarski, A.E. Evolution of public response to the energy crisis. Science, 1974, 184, 257-263.

National Academy of Sciences. Sociopolitical effects of energy use and policy. Reports to the Sociopolitical Effects Resource Group, Risk and Impact Panel of the Committee on Nuclear and Alternative Energy Systems (CONAES), Supporting Paper No.5. Washington, D.C.: National Research Council, 1979.

National Academy of Sciences. Energy choices in a democratic society. The report of the Consumption Location, and Occupational Patterns Resource Group Synthesis Panel of the Committee on Nuclear and Alternative Energy Systems (CONAES), Supporting Paper No. 7. Washington, D.C.: National Research Council, 1980.

National Center for Education Statistics, U.S. Department of Health, Education, and Welfare, Education Division. Energy knowledge and attitudes. A national assessment of energy awareness among young adults. U.S. Department of Health, Education, and Welfare, 1978.

Nelissen, N.J.M. Sociale oorzaken van milieu-onhygiëne; Een onderzoeksnotitie. Sociologische Gids, 1977, 24, 396-406.

Nelson, R.O. Assessment and therapeutic functions of self-monitoring. In: M. Hersen, R.M. Eisler., & P.M. Miller (Eds.), Progress in behavior modification (Vol. 5). New York: Academic Press, 1977.

Neuman, K.A. Correlates of personal resource conservation (first draft). Program in Social Ecology, University of California at Irvine, Irvine, CA: 1980.

Neuman, K.A. Human values: do they make a difference in individuals' commitment to energy conservation? Paper presented at the International Conference on Consumer Behavior and Energy Policy, Noordwijkerhout, September 1982.

Newsom, T.J., & Makranczy, U.J. Reducing electricity consumption of residents living in master metered dormitory complexes. Journal of Environmental Systems, 1977-78, 7, 215-235.

Nietzel, M.T., & Winett, R.A. Demographics, attitudes and behavioral responses to important environmental events. American Journal of Community Psychology, 1977, 5, 195-206.

Nota Energiebeleid. Deel 1/ Algemeen. 's Gravenhage: Staatsuitgeverij, 1979.

Odell, P., & Rosing, K.J. The future of oil: A simulation study of the interrelationships of resources, reserves and use, 1980-2080. London: Kogan Page, 1980.

Olsen, M.E. Consumers' attitudes toward energy conservation. Journal of Social Issues, 1981, 37, 108-131.

Olsen, M.E. Public acceptance of consumer energy conservation strategies. Journal of Economic Psychology, 1983, 183-196.

Olsen, M.E., & Cluett, C. Evaluation of the Seattle City Light neighborhood energy conservation program. Seattle, WA: Batelle Human Affairs Research Centers, 1979.

Olsen, M.E., & Goodnight, J.A. Social aspects of energy conservation. Northwest Energy Policy Project, Study Module 1-B, Final report, 1977.

Olsen, M.E., & Joerges, B. The proces of consumer energy conservation: An international perspective. Paper presented at the Conference on Societal Responses to the Energy Crisis, Dubrovnik, September 1981.

Olson, M. Jr. The logic of collective action: Public goods and the theory of groups. Cambridge, MA: Harvard University Press, 1965.

Olson, M. Jr., & Zeckhauser, R. An economic theory of alliances. Review of Economics and Statistics, 1966, 48, 266-279.

Opzet Nota, Maatschappelijke discussie over de toepassing van kernenergie voor electriciteitsopwekking. Tweede Kamer, zitting 1978-1979, 15100, nr. 18, 1979.

O'Riordan, T. Attitudes, behavior, and environmental policy issues. In: I. Altman & J.F. Wohlwill (Eds.) Human behavior and environment: Advances in theory and research (Vol. I). New York: Plenum, 1976.

Oskamp, S. Psychology's role in the conserving society. Presidential Address to the Division of Population and Environmental Psychology at the American Psychological Association meeting, Los Angeles, 1981.

Otway, H.J., & Fishbein, M. The determinants of attitude formation: An application to nuclear power. Research Memorandum RM-76-80, Austria, Laxenburg: International Institute for Applied Systems Analysis, 1976.

Otway, H.J., & Fishbein, M. Public attitudes and decision making. Research Memorandum RM-77-54, Austria, Laxenburg: International Institute for Applied Systems Analysis, 1977.

Otway, H.J., Maurer, D., & Thomas, K. Nuclear power: The question of public acceptance. Futures, 1978, 10, 109-118.

Palmer, M.H., Lloyd, M.E., & Lloyd, K.E. An experimental analysis of electricity conservation procedures. Journal of Applied Behavior Analysis, 1977, 10, 665-671.

Perlman, R., & Warren, R. Families in the energy crisis: Impacts and implications for theory and policy. Cambridge, MA: Ballinger Publishing Co., 1977.

Philips, P. Household energy consumption attitudes. Unpublished manuscript, Massey University, Department of Geography, New Zealand Energy Research and Development Committee, Report no. 10, Palmerston North, 1976.

Platt, J. Social traps. American Psychologist, 1973, 28, 641-651.

Pomazal, R.J., & Jaccard, J.J. An informational approach to altruistic behavior. Journal of Personality and Social Psychology, 1976, 33, 317-326.

Potma, Th. Het vergeten scenario. Minder energie, toch welvaart. Amsterdam: Meulenhoff Informatief bv., 1979.

Proshansky, H.M., Ittelson, W.H., & Rivlin, L.G. (Eds.). Environmental psychology: People and their physical settings. New York: Holt, Rinehart & Winston, 2nd ed., 1976.

Punter, H. Milieuvriendelijk - Irrationeel. Het milieuprobleem als sociaal dilemma. Groningen: Rijksuniversiteit Groningen, Instituut voor Persoonlijkheidspsychologie, Heymans Bulletins (HB 80-493-EX), 1980.

Raaij, W.F. van. Consumer choice behavior: An information processing approach. Katholieke Hogeschool, Tilburg, 1977. (dissertation).

Raaij, W.F. van. Verspreiding van energiebesparende innovaties onder huishoudens. In: P. Ester & F.L. Leeuw (Eds.), Energie als maatschappelijk probleem. Assen: Van Gorcum, 1981.

Raaij, W.F. van. Micro and macro economic psychology. Paper presented at the 7th Colloquium of Economic Psychology, Edinburgh, July 21-24, 1982.

Raaij, W.F. van., & Eilander, G. Consumentenbezuinigingstactieken. Economisch-Statistische Berichten, 1983, 68, 544-547.

Raaij, W.F. van., & Verhallen, Th.M.M. A behavioral model of residential energy use. Journal of Economic Psychology, 1983, 3, 39-63.

Rappeport, M., & Labaw, P. Trends in energy consumption and attitudes toward the energy shortage. Highlight Report volume V. Princeton, NJ: Opinion Research Corporation, 1974.

Regan, D.T., & Fazio, R.H. On the consistency between attitudes and behavior: Look to the method of attitude formation. Journal of Experimental Social Psychology, 1977, 13, 28-48.

Resources for the Future. Energy in America's future, The choices before us. A study by the staff of the RFF National Energy Strategies Project by S.H. Schurr (project director), J. Darmstadter, H. Perry, W. Ramsay, M. Russell. Baltimore: John Hopkins University Press, 1979.

Richards, C.S. Assessment and behavior modification via self-monitoring: An overview and a bibliography. JSAS Catalog of Selected Documents in Psychology, 1977, 27, 298-313 (Ms. NO. 214).

Ritsema, B.S.M., Midden, C.J.H., & Heijden, P.G.M. van der. Energiebesparing in gezinshuishoudingen: attitudes, normen en gedragingen, een landelijk onderzoek. Werkgroep Energie- en Milieu-onderzoek Rijksuniversiteit Leiden / Energie Studie Centrum - Energieonderzoek Centrum Nederland, 1982.

Robertson, T.S. Innovative behavior and communication. New York: Rinehart and Winston, 1971.

Rogers, E., & Kincaid, D. Communication network analysis: A new paradigm for research. New York: Free Press, 1980.

Rogers, E., & Shoemaker, F. Communication of innovations: A cross-cultural approach. New York: Free Press, 1971.

Ross, M.H., & Williams, R.H. Energy efficiency: Our most underrated resource. The Bulletin of the Atomic Scientists, 1976, November, 30-38.

Rohles, F.H., Jr. Thermal comfort and strategies for energy conservation. Journal of Social Issues, 1981, 37, 132-149.

Rozendal, P.J., Ester, P., & Meer, F. van der. Milieu-attitude als determinant van individueel milieugedrag. Gedrag, 1983, 11, 122-134.

Run, G.J. van., & Wolters, F.J.M. Gedragingen en verwachtingen in dilemma-situaties. Groningen, Rijksuniversiteit Groningen, Instituut voor Persoonlijkheidspsychologie, Heijmans Bulletins (HB 80-451-EX), 1980.

Schelling, T.C. The ecology of micromotives. Public Interest, 1971, 25, 61-98.

Schelling, T.C. Micromotives and macrobehavior. New York: W.W. Norton & Company, 1978.

Schipper, L., & Lichtenberg, A.J. Efficient energy use and well-being: The Swedish example. Science, 1976, 194, 1001-1013.

Schlegel. R.P., Crawford, C.A., & Sanborn, M.D. Correspondence and mediational properties of the Fishbein model: An application to adolescent alcohol use. Journal of Experimental Social Psychology, 1977, 13, 421-430.

Schleyer, W.T., & Young, D.M. Consumer attitudes toward solar energy. (Harvard Business School Energy Project Report). Harvard Business School, 1977.

Schwartz, S.H., & Tessler, R.C. A test of a model for reducing measured attitude-behavior discrepancies. Journal of Personality and Social Psychology, 1972, 24, 225-236.

Schumacher, E.F. Small is beautiful. New York: Harper and Row, 1973.

Sears, D., Tyler, T., Citin, J., & Kinder, D. Political system support and public response to the energy crisis. American Journal of Political Science, 1978, 22, 56-82.

Seaver, W.B., & Patterson, A.H. Decreasing fuel-oil consumption through feedback and social commendation. Journal of Applied Behavior Analysis, 1976, 9, 147-152.

Seligman, C., & Darley, J.M. Feedback as a means of decreasing residential energy consumption. Journal of Applied Psychology, 1977, 62, 363-368.

Seligman, C., & Becker, L.J. (Issue Editors). Energy conservation. Journal of Social Issues, 1981, 37 (entire issue).

Seligman, C., Darley, J.M., & Becker, L.J. Behavioral approaches to residential energy conservation. Energy and Buildings, 1978, 1, 325-337.

Seligman, C., & Hutton, B.R. Evaluating energy conservation programs. Journal of Social Issues, 1981, 37, 51-72.

Seligman, C., Kriss, M., Darley, J.M., Fazio, R.H., Becker, L.J., & Pryor, J.B. Predicting summer energy consumption from homeowners' attitudes. Journal of Applied Social Psychology, 1979, 9, 70-90.

Slavin, R.E., & Wodarski, J.S. Using group contingencies to reduce natural gas consumption in master metered apartments. John Hopkins University, Center for Social Organization of Schools, Report no. 232, Baltimore, 1977.

Slavin, R.E., Wodarski, J.S., & Blackburn, B.L. A group contingency for electricity conservation in matermetered apartments. Journal of Applied Behavior Analysis, 1981, 14, 357-363.

Slovic, P., Fischhoff, B., & Lichtenstein, S. Perception and acceptability of risk from energy systems. in: A. Baum & J.E. Singer (Eds.) Advances in environmental psychology, Vol. 3, Energy: psychological perspectives. Hillsdale, N.J.: Lawrence Erlbaum Associates, 1981.

Socolow, R.H. The Twin Rivers program on energy conservation in housing: Highlights and conclusions. Energy and Buildings, 1978, 1, 207-242.

Socolow, R.H., & Sonderegger, R.C. The Twin Rivers program on energy conservation in housing: Four year summary report (Report no. 32). Princeton, N.J.: Princeton University, Center for Environmental Studies, 1976.

Sonderegger, R.C. Movers and stayers: The resident's contribution to variation across houses in energy consumption for space heating. Energy and Buildings, 1978, 1, 313-324.

Songer-Nocks, E. Situational factors affecting the weighting of predictor components in the Fishbein model. Journal of Experimental Social Psychology, 1976, 12, 56-69.

Sparrow, F.T., Warkow, S., & Kass, R.C. Socioeconomic factors affecting the adoption of solar technology: First findings. In: S. Warkow (Ed.), Energy policy in the United States, Social and behavioral dimensions. New York: Praeger Publishers, 1978.

Stallen, P.J.M., & Meertens, R.W. Beoordeling van risico's van kernenergie. In: P. Ester (Ed.) Sociale aspecten van het milieuvraagstuk. Assen: Van Gorcum, 1979.

Stern, P.C. Effects of incentives and education on resource conservation decisions in a simulated commons dillema. Journal of Personality and Social Psychology, 1976, 34, 1285-1292.

Stern, P.C., Black, J.C., & Elworth, J.T. Personal and contextual influences on household energy adaptations. Paper presented at the International Conference on Consumer Behavior and Energy Policy, Noordwijkerhout, September 1982.

Stern, P.C., Black, J.C., & Elworth, J.T. Response to changing energy conditions among Massachusetts households. Energy, 1983, 8, 515-523.

Stern, P.C., & Gardner, G.T. A review and critique of energy research in psychology. Social Science Energy Review, 1980, 3 (entire issue), Yale University, New Haven.

Stern, P.C., & Kirkpatrick, E.M. Energy behavior. Environment, 1977, 19, 10-15.

Stobaugh, R., & Yergin, D. Energy future: Report of the energy project of the Harvard Business School. New York: Random House, 1979.

Stokols, D. Environmental psychology. Annual Review of Psychology, 1978, 29, 253-295.

Stutzman, T.M., & Green, S.B. Factors affecting energy consumption: Two field tests of the Fishbein-Ajzen model. Journal of Social Psychology, 1982, 117, 183-201.

Talarzyk, W.W., & Omura, G.S. Consumer attitudes toward and perceptions of the energy crisis. American Marketing Association 1974 Combined Proceedings. Ann Arbor, MI: University Microfilms, 1975.

Tharp, R.G., & Wetzel, R.J. Behavior modification in the natural environment. New York: Academic Press, 1969.

Thomas, K., Maurer, D., Fishbein, M., Otway, H.J., Hinkle, R., & Simpson, D. A comparative study of public beliefs about energy systems. Research Memorandum RM-78-XX, Austria, Laxenburg: International Institute for Applied Systems Analysis, 1978.

Thompson, P.T., & MacTavish, J. Energy problems: public beliefs, attitudes and behaviors. Unpublished manuscript, Grand Valley State Colleges, Urban and Environmental Studies Institute, Allendale: Michigan, 1976.

Thoresen, C.E., & Mahoney, M.J. Behavioral self-control. New York: Holt, Rinehart & Winston, 1974.

Tuso, M.A., & Geller, E.S. Behavior applied to environmental/ecological problems: A review. Journal of Applied Behavior Analysis, 1976, 9, 526.

Ultee, W.C. Groei van kennis en stagnatie in de sociologie. Utrecht: Rijksuniversiteit, 1977. (dissertation).

Van Liere, K.D., & Dunlap, R.E. The social bases of environmental concern: A review of hypotheses, explanations and empirical evidence. Paper presented at the Annual Meeting of the Southern Sociological Society, Atlanta, 1979.

Verhallen, Th.M.M., & Raaij, W.F. van. Household behavior and energy consumption. Tilburg University, Department of Psychology, Tilburg, 1979.

Verhallen, Th.M.M. & Raaij, W.F. van. Household behavior and energy use. Paper presented at the International Conference on Consumer Behavior and Energy Use, Banff, Alberta, September 1980.

Verhallen, Th.M.M., & Raaij, W.F. van. Household behavior and the use of natural gas for home heating. The Journal of Consumer Research, 1981, 8, 253-257.

Visser, P.E. Collectieve doelstelling en individueel gedrag: de energiebesparing. Maandschrift Economie, 1980, 44, 220-228.

WAES (Workshop on Alternative Energy Strategies). Energy: global prospects 1985-2000. New York: McGraw-Hill Book Company, 1977.

Walker, J.M. Energy demand behavior in a master-metered apartment complex. An experimental analysis. Journal of Applied Psychology, 1979, 64, 190-196.

Walker, N.E., & Draper, E.C. The effects of electricity price increases on residential usage of three economic groups: A case study. In: Texas nuclear power policies, Vol. 5: Sociodemographic and economic effects. Center for Energy Studies, University of Texas, Austin, TX, 1975.

Warkov, S. (Ed.). Energy policy in the United States: Social and behavioral dimensions. New York: Praeger Press, 1978.

Warkov, S. Energy conservation and adoption of household solar. Paper presented at the International Conference on Consumer Behavior and Energy Use, Banff, Alberta, September 1980.

Warren, D.I., & Clifford, D.L. Local neighborhood social structure and response to the energy crisis of 1973-73. Institute of Labor and Industrial Relations, University of Michigan, Ann Arbor, MI, 1975.

Weigel, R.H., Vernon, D.T.A., & Tognacci, L.N. Specificity of the attitude as a determinant of attitude-behavior congruence. Journal of Personality and Social Psychology, 1974, 30, 724-728.

Weiss, P.R. Public attitudes and solar policy. Golden, Co.: Solar Energy Research Institute, 1979. Report prepared for the U.S. Department of Energy. SERI/TP-51-188.

Wilbanks, T.J. Introduction to K.M. Gentemann (Ed.). Social and political perspectives on energy policy. New York: Praeger Press, 1981.

Wilson, J.A. A test of the Tragedy of the Commons. In: G. Hardin & J. Baden (Eds.) Managing the commons. San Francisco, CA: W.H. Freeman and Company, 1977.

Winett, R.A. An emerging approach to energy conservation. In: D. Glenwick & I. Jason (Eds.). Behavioral community psychology, progress and prospects. New York: Praeger Press, 1980.

Winett, R.A., & Ester, P. Behavioral science and energy conservation: Conceptualizations, strategies, outcomes, energy policy applications. Journal of Economic Psychology, 1983, 203-229.

Winett, R.A., Hatcher, J.W., Fort, T.R., Leckliter. I.N., Love, S.Q., Riley, A.W., & Fishback, J.F. The effects of videotape modeling and daily feedback on residential electricity conservation, home temperature and humidity, perceived comfort, and clothing worn: Winter and summer. Journal of Applied Behavior Analysis, 1982, 15, 381-402.

Winett, R.A., Kagel, J.H., Batalio, R.C., & Winkler, R.C. Effects of monetary rebates, feedback, and information on residential electricity conservation. Journal of Applied Psychology, 1978, 63, 73-80.

Winett, R.A., Kaiser, S., & Haberkorn, G. The effects of monetary rebates and feedback on electricity conservation. Journal of Environmental Systems, 1977, 6, 329-341.

Winett, R.A., Leckliter, I.N., Love, S.A., Chinn, D.E., & Stahl, B. The effects of videotape modeling in group and home settings and via cable TV on residential energy conservation, home temperature and humidity, comfort, and clothing worn: Winter and summer replications and extensions of findings. Manuscript submitted for publication, 1983.

Winett, R.A., Neale, M.S., & Grier, H.C. Effects of self-monitoring and feedback on residential electricity consumption. Journal of Applied Behavior Analysis, 1979, 12, 173-184.

Winett, R.A., Neale, M.S., Williams, K., Yokley, J., & Kauder, H. The effects of feedback on residential electricity consumption: Three replications. Journal of Environmental Systems, 1978, 8, 217-233.

Winett, R.A., & Nietzel, M.T. Behavioral ecology: Contingency management of consumer energy use. American Journal of Community Psychology, 1975, 3, 123-133.

Winkler, R.C., & Winett, R.A. Behavioral interventions in resource conservation. A systems approach based on behavioral economics. American Psychologist, 1982, 37, 421-435.

Wippler, R. Individual action and social outcome: One problem in different traditions. Paper presented at the Conference on Institutions, Individuals, and Collective Action, Utrecht, December 1977.

Woerkum, C.M.J. van. Voorlichtingskunde en massacommunicatie, het werkplan van de massamediale voorlichting. Wageningen: Landbouwhogeschool, 1982. (dissertation).

Woo, T.O., & Castore, C.H. Expectancy-value and selective exposure as determinants of attitudes towards a nuclear power plant. Journal of Applied Social Psychology, 1980, 10, 224-234.

World Energy Conference. World Energy resources 1985-2020. Coal resources, an appraisal of world coal resources and their future availability. Guilford: I.P.C. Press, 1978.

Wotaki, T. The Princeton omnibus experiment: Some effects of retrofits on heating requirements. Princeton University, Center for Environmental Studies, Report No. 43, 1977.

Wrightsman, J., O'Connor, J., & Baker, N. (Eds.). Cooperation and competition: Readings on mixed-motive games. San Francisco: Freeman, 1972.

Yergin, D. Conservation: The key energy source. In: R. Stobaugh & D. Yergin (Eds.) Energy future: Report of the Energy Project at the Harvard Business School. New York: Random House, 1979.

Zuiches, J.J. Acceptability of energy policies to Mid-Michigan families. Unpublished manuscript, Michigan State University, Agricultural Experiment Station, Research Report no. 1976, East Lansing, Michigan, 1976.

AUTHOR INDEX

About the author

Peter Ester studied sociology at the State University of Utrecht and received his Ph.D. from the Erasmus University of Rotterdam. From 1976 to 1980, he was research associate at the Institute for Environmental Studies, Free University of Amsterdam, where he specialized in social aspects of environmental and energy issues. From 1980 to 1984 he was head of the Social Science Department of this research institute. He has published over forty articles and research papers, some fifteen of which appeared in international scientific journals. Dr. Ester is co-author/editor of four books and associate editor of Environment and Behavior as well as of the International Handbook of Environmental Psychology (John Wiley & Sons, Inc., in press).

In 1981 he was associated as a Fulbright scholar with the Department of Psychology, Claremont Graduate School, California and the Department of Psychology, Virginia Polytechnic Institute and State University, Blacksburg, Virginia.

Since 1984 he is in charge of the Department of Social Reporting and Advice of the Dutch Social and Cultural Planning Office (Rijswijk), which conducts social research in the areas of education, labor, social security, media, leisure time, innovations, emancipation, cultural changes, ethnic minorities, crime and justice, public health, social services, public housing, physical planning, and social indicators.